Electronegativity

Nazmul Islam

Electronegativity

The concepts, scales and applications of electronegativity

LAP LAMBERT Academic Publishing

Impressum/Imprint (nur für Deutschland/ only for Germany)
Bibliografische Information der Deutschen Nationalbibliothek: Die Deutsche Nationalbibliothek verzeichnet diese Publikation in der Deutschen Nationalbibliografie; detaillierte bibliografische Daten sind im Internet über http://dnb.d-nb.de abrufbar.

Coverbild: www.ingimage.com

Verlag: LAP LAMBERT Academic Publishing GmbH & Co. KG
Dudweiler Landstr. 99, 66123 Saarbrücken, Deutschland
Telefon +49 681 3720-310, Telefax +49 681 3720-3109
Email: info@lap-publishing.com

Herstellung in Deutschland:
Schaltungsdienst Lange o.H.G., Berlin
Books on Demand GmbH, Norderstedt
Reha GmbH, Saarbrücken
Amazon Distribution GmbH, Leipzig
ISBN: 978-3-8383-7949-4

Imprint (only for USA, GB)
Bibliographic information published by the Deutsche Nationalbibliothek: The Deutsche Nationalbibliothek lists this publication in the Deutsche Nationalbibliografie; detailed bibliographic data are available in the Internet at http://dnb.d-nb.de.

Cover image: www.ingimage.com

Publisher: LAP LAMBERT Academic Publishing GmbH & Co. KG
Dudweiler Landstr. 99, 66123 Saarbrücken, Germany
Phone +49 681 3720-310, Fax +49 681 3720-3109
Email: info@lap-publishing.com

Printed in the U.S.A.
Printed in the U.K. by (see last page)
ISBN: 978-3-8383-7949-4

Electronegativity

Nazmul Islam
Department of Basic Sciences & Humanities,
Techno Global-Balurgaht,
Balurghat- 733101
Email- nazmul.islam786@gmail.com
nazmul.islam@rediffmail.com
nazmul.islamtgbg@gmail.com

About Author

Nazmul Islam submitted his PhD thesis in 2010 and already made good number of publications in various prestigious international peer reviewed journals.He is currently a Lecturer at the Techno Global-Balurgaht. Acted as a Coordinator,visiting faculty,and counselor of various PG and UG Colleges.

Acknowledgement

I would like to thank my post graduate student Mr. Chandra Chur Ghosh for his help in this work

Abstract

This work is composed of three parts.

In the Chapter -1, a systematic account of the concepts and the trends in the development of electronegativity, a fundamental property of atoms, ions, groups, bonds and molecules, is given. Attention is concentrated to the necessity of taking into account the characteristic features of the electronegativity. The problems and the modifications of various scales of electronegativity are discussed.

In the Chapter-2 of the project, we have reviewed some important applications of the concept of electronegativity in the chemical world.

In the previous parts, we have critically analyzed the scales and concepts electronegativity, but the electronegativity is a conundrum and occurs in mind. It is neither a physical observable nor a quantum mechanically determinable quantity and for this reason there is no benchmark to perform any validity test of any scale of electronegativity. But the concept of electronegativity has extremely useful applications in the real world. In the Chapter-2 of the present work, we analyzed some important applications of electronegativity viz, the bond energies, bond polarities and the dipole moments, force constants, internuclear distance, standard enthalpy, atomic polar tensor etc. In the Chapter-3, we have calculated the molecular electronegativity, internuclear distance, atomic polar tensor (APT), bond energy, standard enthalpies of formation, dipole charge, and dipole moment of some extremely ionic compounds and covalent molecules. We have found some dimensional inconsistency in the algorithms for the evaluation of bond energy. We venture to suggests a new semi-empirical ansatz for the evaluation of the bond energy(BE) as $BE=a(\Delta\chi/R_{AB})+b$ in energy unit, where $\Delta\chi$ is the electronegativity difference and R_{AB} is the internuclear distance of the constituents forming the bond. We have applied our suggested ansatz to calculate the bond energy of the alkali halide and hydrogen halide molecules.

We also made comparative study of the evaluated bond energy of some selected molecules *vis a vis* their experimental counterparts and found beautiful correlation between them. Furthermore, we proposed a new formula for the evaluation of standard enthalpies of formation as $-H_f=a\Delta\chi^2+b$. We have performed comparative studies of the data for the standard enthalpies of formation evaluated through the proposed ansatz, $-H_f=a\Delta\chi^2+b$, of alkali halides vis a vis their experimental counterparts and have found nice correlation. The comparative study of the dipole charge evaluated using different algorithms of alkali halides

The detail study concludes that electronegativity is an intrinsic property of atoms, not in situ property which originates from the attraction of the screened nucleus of the atom on the valence electron. It is important fundamental descriptors which has no quantum mechanical observable, hence empirical. There is no experimental benchmark for electronegativity as it is not experimental determinable quantity. Electronegativity has wide application in the real world of chemical interaction and chemical stability.

CHAPTER-1

Concepts and Scales of electronegativity

Index

2.10. Yuan scale of electronegativity (1964):

2.11. Gyftopoulos and Hatsopoulos quantum thermodynamic scale of electronegativity (1968):

2.12. Phillips scale of electronegativity (1968):

2.13. St. John and Bloch quantum defect scale of electronegativity (1974):

2.14. Density functional scales of Electronegativity (1978):

2.14.1 Parr electronegativity scale (1978):

2.14.2 Parr and Pearson finite difference scale of electronegativity (1983):

2.14.3 Pearson frontier orbital scale of electronegativity (1986):

2.15. Pasternak scale of electronegativity (1978):

2.16. Zhang scale of electronegativity (1982):

2.17. Boyd and Edgecombe scale of electronegativity (1988):

2.18. Allen scale of electronegativity (1989):

2.19. Nagle scale of electronegativity (1990):

2.20. Zheng and Li scale of electronegativity (1990):

2.21. Ghosh scale of electronegativity (2005):

1.Introduction:

Electronegativity is one of the most enduring and important tools of science and an animated field of current research. The concept of electronegativity is very old and useful in the field of fundamental science purportedly introduced two centuries ago by Avogadro [1].

The concept of electronegativity was instigated in 1811 when Avogadro [1] pointed out the correlations between the neutralization that occurs with acids and bases, and the neutralization that occurs between positive and negative electrical charges. Avogadro proposed that these cancellation relationships could be applied not only to the simple substances but also to the more complex compounds. From his own experiences and also by the earlier works of Davy and Volta [2], and Heinrich Pfaff [3], Avogadro proposed an “oxygenicity scale” on which every element could be placed depending upon its tendency to react with other elements. This scale is believed as the forerunner of the modern electronegativity scale. Jöns Jacob Berzelius [5] worked further with the fundamental fact that oxygen, acid, and oxidized substances accumulate in the region of the positive pole of an electrolytic cell, whereas metals, bases, and combustible substances accumulate around the negative pole. Introducing the concept that caloric was created by the combination of negative and positive electricity, Berzelius modified Avogadro’s oxogenicity scale. Berzelius coined the term “electronegativity” instead of "oxygenicity" and categorized elements into two classes-(i) electronegative and (ii) electropositive. Assuming both heat and electricity as fluids, he formulated a 'universal scale of electronegativity' of the elements. Latter it was found that the electronegativities of elements measured by Berzelius coordinate remarkably well with the thermo chemical scale of Pauling [6] and also the force concept scale of Allred and Rochow [7].

But, Berzelius theory failed to account for half of all possible chemical reactions such as endothermic associations and exothermic dissociations. Some other drawbacks of this theory were it could not account for increasingly complex organic molecules, and was incompatible with Michael Faraday's laws of electrolysis.

The scientific world had to wait till the publication of the landmark work, "The Nature of Chemical Bond" onto which based on thermo chemical data, the definition and the scale of electronegativity was introduced by Linus Pauling [6].

Pauling first gave the objection for the use of electrode potential as a measure of electron attracting power. We quoted the original from Pauling "*the property of the electronegativity of an atom is different from the electrode potential of the element which depends on the difference in free energy of the element in its standard state and in ionic solution, and it is different from the ionization potential of the atom and from the electron affinity; although it is related to the properties in a general way*".

Lewis, in 1916, advanced an idea that the electron pair connecting two dissimilar atoms is attracted to one of the atom more strongly [6]. According to him, in the general case a covalent bond formed between two dissimilar atoms, has partially ionic character due to the displacement of the bonding electrons to one of the atoms. Keeping this suggestion in mind, and based on thermodynamical and quantummechanical arguments, Pauling defined electronegativity as '*the power of an atom in a molecule to attract electron pair toward itself*'. In other word, following Pauling we may define electronegativity as the property of atoms which governs the electron distribution of an atom in a molecule.

Till then, the concept coined by Pauling, electronegativity has become an indispensable tool in every branch of chemistry, physics and biology. It is very essential tool for both theorists and experimentalists. So every student of chemistry, physics and biology must have the proper knowledge of the concepts, scales and application of electronegativity. We may site come comments of various doyens of chemistry and physics from which one can easily understand the significance of electronegativity. Allen [8] considered electronegativity as the configuration energy of the system and argued that electronegativity is a fundamental atomic property and is the missing third dimension to the periodic table. He further assigned electronegativity as an 'ad hoc' parameter. Huheey [9] opined that the concept of electronegativity is simultaneously one of the most important and difficult problems in chemistry. Ghosh and Islam [10] recently opined that the appearance and the significance of the concepts of electronegativity (and also the hardness) in chemistry and physics resemble the “unicorns of mythical saga”, which has no physical sense but without the concept and operational significance of which chemistry becomes disordered and the long established unique order in chemico-physical world will be taken aback.

Electronegativity is a fundamental descriptor of atoms molecules and ions which can be used in correlating a vast field of chemical knowledge and experience. e.g, the static and dynamic behavior of molecules can be well understood [11] by the use of the electronegativity concept. The fundamental quantities of inorganic, organic, and physical chemistry such as bond energy, polarity, and the inductive effect can be visualized in terms of electronegativity. At present, the concept of electronegativity is not only widely used in chemistry but also in biology, physics and geology [12, 13]. An outstanding dependence of the superconducting transition temperature on electronegativity is found for both solid elements and high-temperature superconductors [14-16]. Electronegativity concept has been successfully used to correlate various spectroscopic phenomenons such as nuclear quadruple coupling from microwave and radio wave frequency spectroscopy [17] and with the chemical shift in nuclear magnetic resonance spectroscopy [18]. Lackner and Zweig [19] pointed out that the electronegativity has led to the correlation of vast number of important atomic and molecular properties and also to qualitative understanding of quark atoms. The concept of electronegativity has been successively used by Correia et al. [20] Hur et al [22] and Zhang [23] to explain the geometry and properties of molecule such as superconductivity, photocatalytic activity, magnetic property and optical basicity. In recent years, electronegativity concept has been used to material design [24], drug design [25] etc. The manifold application of electronegativity draws the attention of students of chemistry, physics and biology over more than a century. It is in this spirit that electronegativity is re-examined in this work.

2. Various scales of electronegativity:

Ever since the concept of electronegativity was presented by Pauling, many attempts have been made to quantify this concept through the establishment of various electronegativity scales. But till now, scientific world believe that the final scale of electronegativity is not proposed by any one. Electronegativity is empirical and will empirical as there is no quantum mechanical operator for it and also electronegativity is not experimentally measurable quantity [10]. In this section, we reviewed some of the most important and useful scales of electronegativity of atoms, ions and orbitals. Attention is concentrated mainly on the atomic electronegativity scales, but some important orbital and ionic scales related to the atomic electronegativity are also discussed.

2.1. Pauling thermo-chemical scale of electronegativity (1932):

In the first edition of "The Nature of the Chemical Bond", Pauling [6] by an ingenious mixing of thermodynamical and quantummechanical arguments presented the meaning of the word "electronegativity" as *"the power of an atom in a molecule to attract electrons toward itself."* Pauling's empirical thermo chemical scale of electronegativity is based on heats of formation of covalent bonds.

Pauling forwarded the concept that hybridized atomic orbitals (AO) are the distinguishing factor of the valence state of an atom. A covalent bond A-B, can be well described by the localized molecular orbital (LMO) of the bonded atoms. Coupling of electrons to form X-Y bond is accompanied by the intervention of the electron waves and also the charge transfer from the less electronegative atom to the more electronegative atom. Charge transfer is responsible for ionic character of a bond and is characterized by difference in the population of AOs of the associated atoms X and Y in the molecule XY. On the basis of the above qualitative idea, Pauling supposed that the energy of an ordinary covalent bond X-Y is generally larger than the additive mean of the energies of the bond X-X and Y-Y and the enhancement factor Δ which increases as the atoms X and Y become more and more unlike in their electronegativity property. Considering the electronegativities of X and Y are χ_X and χ_Y, Pauling proposed the relationship between the electronegativity difference and the enhancement factor as

$$\chi_X \sim \chi_Y = 0.208\sqrt{\Delta} \qquad (1)$$

The enhancement factor Δ calculated by him as

$$\Delta = D_{(X\text{-}Y)} - 0.5[D_{(X\text{-}X)} + D_{(Y\text{-}Y)}] \qquad (2)$$

where the dissociation energies, D, of the X–Y, X–X and Y–Y bonds are expressed in eV unit.

As only differences in electronegativity are defined in Pauling scale, it was necessary to choose an arbitrary reference point in order to construct a scale. Hydrogen was chosen as the reference by Pauling, as it forms covalent bonds with a large variety of elements and its electronegativity was fixed as 2.1.

The unit of electronegativity in Pauling scale is $(\text{energy})^{1/2}$. Now this unit is referred as thermochemical unit (TU) [26]

Pauling [6] proposed electronegativity values for thirty-three elements. There after a number of workers revisited and extended Pauling scale in order to quantify Pauling's electronegativity concept.

Haissinsky [27] in 1946 extended Pauling's calculations to seventy-three elements. The author also showed that for multivalent elements electronegativity is a function of valency of the atoms. In 1953, Huggins [28] re-evaluated the electronegativities of seventeen elements of the periodic table. Gordy and Orville Thomas [29] pointed out that the Huggins' electronegativity values are generally higher than Pauling electronegativity values. They opined further that if the Huggins' electronegativity values are downgrade by 0.1 and the values are round off to two significant figures than the Huggins' electronegativity values agree well to that of Pauling values.

Altshuller [30] evaluated electronegativity values of the copper, Zinc and Gallium sub group elements. Thereafter, in 1961, Allred [31] revisited the Pauling electronegativity scale and calculated electronegativities of sixty-nine elements from the thermochemical data published at that time. He summarized the trends of electronegativity values within the periodic system.

Introducing LCAO coefficient in the electronegativity concept, a theoretical basis of Pauling scale was suggested by Mulliken [32].

It is apparent from Pauling definition that electronegativity is not the property of isolated atom but it depends on the molecular environment in which the atom is present.

2.2. Malone scale of Electronegativity (1933):

In the year 1933, Malone [33] suggested a relationship between the dipole moment in Debye (μ_d) of a hetaronuclear bond X-Y and the electronegativity difference, χ_X~χ_Y, as-

$$\chi_X \sim \chi_Y \propto \mu_d \qquad (3)$$

Malone's scale of electronegativity can be applied remarkably well in a few well known cases (e.g. hydrogen halides) but cannot be applied in case of others compounds because of the inaccuracy in the result, and hence cannot be accepted as a reliable measure.

2.3. Mulliken Scale of Electronegativity (1934):

One of the most pre-eminent electronegativity scale was created by Mulliken [34] in 1934. Mulliken opined that the arithmetic mean of the first ionization energy and the electron affinity of an atom can be used as representative of the tendency of an atom to attract electrons and hence it represents the electronegativity of the atom. Thus the empirical spectroscopic definition of electronegativity was suggested as the average of the ionization potential (IP) and electron affinity (EA) for the valence state of an atom.

Let us have a look on the work of Mulliken.

Mulliken considered two limiting resonance structures of the diatomic complex XY.

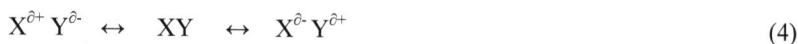

$$X^{\partial+} Y^{\partial-} \leftrightarrow XY \leftrightarrow X^{\partial-} Y^{\partial+} \qquad (4)$$

If one replaces the limiting structures by the equivalent ionic components then equation (4) looks like this-

$$X^{+} + Y^{-} \leftrightarrow X + Y \leftrightarrow X^{-} + Y^{+} \qquad (5)$$

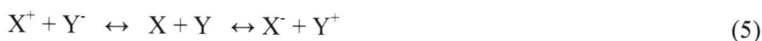

Case-1: Y is more electronegative than X, Y holds more negative charge than X ie, left hand side of the reaction (4) and reaction (5).

In this case,

$$X + Y \rightarrow X^{+} + Y^{-} \qquad (6)$$

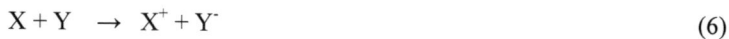

Energy change associated with the reaction (6) is given by the difference between the energy required to remove an electron from A, its ionization potential (I), and the energy consumed to attach the electron to the outer shell of B, its electron affinity (A)

$$\Delta E_{(X^{+}Y^{-})} = I_X - A_Y \qquad (7)$$

Case-2: X is more electronegative than Y, X hold more negative charge than Y ie, right hand side of the reaction (4) and reaction (5).

Similarly, for that the other equivalent reaction,

$$X + Y \rightarrow X^{-} + Y^{+} \quad (8)$$

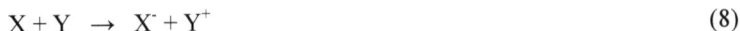

the consumed energy is

$$\Delta E_{(X^- Y^+)} = I_Y - A_X \quad (9)$$

Now, Mulliken's assumption was that the difference between A+B- and A-B+ can be neglected as they are not truly ionic. So the involved energies, $\Delta E_{(A^+ B^-)}$, $\Delta E_{(A^- B^+)}$ can be equalized.

$$\Delta E_{(X^+ Y^-)} = \Delta E_{(X^- Y^+)} \quad (10)$$

$$\text{i.e, } I_X - A_Y = I_Y - A_X \quad (11)$$

$$\text{or, } I_X + A_X = I_Y + A_Y \quad (12)$$

Equation (12) reveals that the sum of ionization energy and electron affinity of each separate atom becomes equal when they are combined to form the complex, XY.

Previously, Hund [35] stated that the quantities average of I and A, i.e., (I+A)/2, is an approximate criterion for their equal participation in a chemical bond.

From this idea, Mulliken considered an arithmetic mean of the first ionization energy and electron affinity as a qualitative definition of electronegativity for any species X (atom, molecule, or radical in its state of interaction)-

$\chi_X \approx (I_X + A_X)/2$ (13)

As this definition is not dependent on an arbitrary relative scale, it has also been termed absolute electronegativity, with the units of kilojoules per mole or electron volts.

However, it is more usual to use a linear transformation to transform these absolute values into values which resemble the more familiar Pauling values. Scaling with Pauling electronegativity values, the Mulliken electronegativity was expressed as

$\chi = a\ (I+A) + b$ (14)

if I and A are expressed in electron volt then

$\chi = 0.187(I+A) + 0.17$

Coulson [36] remarked that Mulliken's measure of electronegativity is better and more precise than that of Pauling's scale of electronegativity.

Mulliken scale is absolute and more fundamental because it depends on the fundamental energy value for isolated atom. It is important because it has bearing the modern density functional definition of electronegativity [37].

$\chi_{DFT} = -(\partial E/\partial N)v$ (15)

From the energy versus number of electron curve, it is transparent that the change in energy, ΔE, is associated with two electrons changes. If we consider S as a neutral species having energy E_N, and having a total number of N electrons, then the corresponding cation and anion, S^+ and S^- have the energy E_{N-1} and E_{N+1} and the number of electrons N-1 and N+1 respectively. Then the energy change and the change in number of electrons can simply be calculated using the following equations

$\Delta E = E_{N+1} - E_{N-1}$ (16)

and $\Delta N = (N+1)-(N-1)=2$

Thus the density functional electronegativity

$\chi_{DFT} = -(\partial E/\partial N)v = -(E_{N+1} - E_{N-1})/2 = \{(E_{N-1}-E_N)+(E_N - E_{N+1})\}/2 = \frac{1}{2}(I+A)$

or,

$\chi_{DFT} = \frac{1}{2}(I+A) = \chi_{Mulliken}$ (17)

Bratsch[38] reviewed the theoretical basis, concept and application of Mulliken electronegativity in terms of valence state promotional energies. He considered the valence state ionizational potential (I_v) and electron affinity (A_v) proposed by Hinze and Jaffe [39] as-

$I_v = I + P^+ - P^0$ (18)

and,

$A_v = A + P^0 - P^-$ (19)

where P stands for valence shell promotional energy.

The terms, 'a' and 'b', was previously defined by Huheey [40] as-

$a=(I_v+A_v)/2=(I_G+A_G+ P^+-P^-)/2$ (20)

$b= I_v-A_v=(I_G-A_G+ P^+-2P^0 +P^-$ (21)

Bratsch[38] showed that the Mulliken 'a' and 'b' parameter for a given element vary linearly with the increasing degree of s character.

Bratsch[38] opined further that a linear relationship between Mulliken and Pauling electronegativity is not possible to propose because of the dimension difference in the two scales. Mulliken electronegativity has the dimension of energy while the Pauling scale has the dimension of the square root of energy. Thus he proposed a linear relationship between the Pauling electronegativity and square root of Mulliken electronegativity as follows-

$\chi_{Pauling} = 1.35(\chi_{Mulliken})^{1/2}-1.37$ (22)

Bratsch[38] calculated partial ionic charge, bond energy and group electronegativity from Mulliken electronegativity and also correlated the Mulliken 'a' and 'b' parameter with the Hard Soft Acid Base Principle [41].

2.4. Gordy Scale of electronegativity (1946):

In 1946 , Gordy[38] suggested that the electronegativity of an atom(χ) is equal to the electrostatic potential(or the effective nuclear charge, Z_{eff}, of the nucleus on the outermost electron) felt by one of its valence electrons at a radial distance equal to atom's single bond covalent radius(r_{cov}). Justifiably such potential created by the conjoint action of the nucleus and the remaining electrons. Thus,

$$\chi_{Gordy} = e\, Z_{eff}/r_{cov} \qquad (23)$$

Unit of electronegativity in Gordy scale is energy per electron.

The electrostatic electronegativity scale of Gordy [38] was scientifically justified by a good number of workers viz. Pasternak [42], Ray, Samuels and Parr [43], Politzer, Parr and Murphy [44] and this scale then accepted in the scientific world widely.

Gordy and Orville Thomas [29] pointed out that the electronegativity ansatz of Gordy cannot be used to calculate the electronegativity values for the transition elements for which the energy levels of different shells begin to overlap. They explained the fact for the deviation and modified the scale. They postulated that the effective nuclear charge Z_{eff} can be obtained with the approximation that all electrons in closed shells below the valence shells use their full screening power and that all valence electrons exert equal screening.

Thus,

$$(Z_{eff})_{Gordy\text{-}Thomas} \approx n - \sigma(n-1) \qquad (24)$$

where n is the number of electron in the valence shell of the atom, σ is the screening constant of the valence electrons.

Substituting Z_{eff} by $(Z_{eff})_{Gordy\text{-}Thomas}$ in equation (23) the electronegativity ansatz proposed by them as follows-

$$\chi_{Gordy\text{-}Thomas} = e\{n - \sigma (n-1)\}/r_{cov} \qquad (25)$$

Recently, Ghosh and Chakraborty [45] pointed out that r_{cov} can not be a necessary input in computing electronegativity as electrostatic potential and they modified Gordy's formula of measuring electronegativity by substituting covalent radii by absolute radii. They also proposed that the electronegativity, χ, is proportional to Z_{eff}/r. Thus the modified Gordy's electronegativity ansatz (23) is-

$$\chi_{Ghosh\ and\ Chakraborty} = a\,(Z_{eff}/r_{abs}) + b \qquad (26)$$

Where 'a' and 'b' are the constants for a given period.

2.5. Walsh scale of electronegativity (1951):

Walsh [46] proposed that the electronegativity of an atom or any group 'X' is the stretching force constants of its bonds to a hydrogen atom (X-H).

2.6. Sanderson scale of electronegativity (1952):

Sanderson [47] noted the relationship between electronegativity and atomic size, and has proposed a method of calculation of electronegativity based on the reciprocal of the atomic volume. With knowledge of bond lengths, Sanderson's method allows to estimate the bond energies in a wide range of compounds.

Sanderson further postulated that when two or more atoms combine to form a molecule, their electronegativities get equalized. This established a new basis for understanding and application of electronegativity.

2.7. Allred and Rochow Scale of electronegativity (1958):

In 1958, Allred and Rochow [7] postulated that electronegativity of an atom is related to the charge experienced by an electron on the outermost shell of an atom, or with the electrostatic field. The higher the charge of atomic surfaces per unit area, the greater is the tendency of that atom to attract electrons.

Now, the charge experienced by an electron on the surface of an atom or on the outermost shell of an atom can be described in terms of the effective nuclear charge, Z_{eff} experienced by valence electrons and the surface area of the atom. Now as the surface area of an atom is proportional to the covalent radius of the atom, the electronegativity can be presented as

$$\chi \propto Z_{eff}/r^2_{cov} \qquad (27)$$

Allred and Rochow suggested a linear relationship between χ and Z_{eff}/r^2_{cov} as,

$$\chi = a(Z_{eff}/r^2_{cov}) + b \qquad (28)$$

Scaling with Pauling [6] electronegativity values, Allred and Rochow [7] proposed a new electronegativity scale as-

$$\chi_{Allred\text{-}Rochow} = 0.359(Z_{eff}/r^2_{cov}) + 0.744 \qquad (29)$$

The concept of Allred and Rochow was justified by a good number of workers viz, Little and Jones [48], Mande et al [49].

Little and Jones [48] verified and recalculated the electronegativity of the atoms of the periodic table based on the force concept of Allred and Rochow.

Mande et.al[49] on the basis of relativistic Dirac equation, calculated screening constants using x-ray spectroscopic method and using the spectroscopic effective nuclear charge of the atoms for the valence states they evaluated the electronegativity of the atoms of the periodic table by the following modified ansatz

$$\chi = 0.778\ (Z_{eff})_{spectroscopic}/r^2_{cov}) + 0.5 \qquad (30)$$

The constants of the above ansatz (30) was evaluated by Mande et al. by plotting $(Z_{eff})_{spectroscopic}/r^2_{cov}$ with the Pauling's electronegativity values.

The electrostatic scale of Allred and Rochow was further modified by Boyd and Markus [50]. They proposed a non empirical electrostatic model for calculating the attraction between the Z_{eff} and an electron at a distance corresponding to the relative radius of the atom. They proposed a non empirical electrostatic electronegativity scale of Allred and Rochow by using some modifications as-

$$\chi_{\text{Boyd-Markus}} = \frac{kZ}{r^2}\left[1 - \int_0^r D(r)\,dr\right] \qquad (31)$$

where, Z is the atomic number, r is the relative radius of the atom, D(r) is the radial density function and k is a constant so chosen that the electronegativity value of F becomes 4. In atomic unit, they assigned the value for the constant k as 69.4793. The relative radius of atom was calculated by them using the density contour approach of Boyd [51] on the basis of analytical Hartree- Fock wave function of atoms proposed by Clementi and Roetti [52].

In a recent work, Ghosh and Chakraborty [53] pointed out some inconsistency in the Allred and Rochow electronegativity scale and also in the modified Allred and Rochow scales by Mande et al. and Little and Jones.

They found that-

i) Although, Allred and Rochow, Mande et al. and Little and Jones used the force concept to evaluate the electronegativity of atoms but the dimension of the electronegativity is not be mentioned by any of them.

ii) All of them used the covalent radii in atomic unit to calculate the electronegativity values in their proposed electronegativity scale.

This does not produce the electronegativity in force unit.

iii) The absolute radius is the true descriptor of atomic electronegativity not the covalent radius.

Ghosh and Chakraborty[53] replaced the covalent radii by absolute radii and solved the dimension problem of the Allred and Rochow scale by proper dimension matching and they reported electronegativity as force. They reported the electronegativity values of 103 elements of the periodic table by the modified Allred and Rochow electronegativity scale.

2.8. Iczkowski and Margrave scale of electronegativity (1961):

Iczkowski and Margrave [54] considered electronegativity as a property of an isolated gaseous atom or ion and they [54] have plotted the atomic energy change with degree of ionization (figure-1) and found that the following expression for the energy of an atom, X as-

$$E(N)_X = aN + bN^2 + cN^3 + dN^4 + \ldots\ldots\ldots \quad (32)$$

where N is the number of electrons in the valence shell of nucleus X, and a,b,c and d are coefficients.

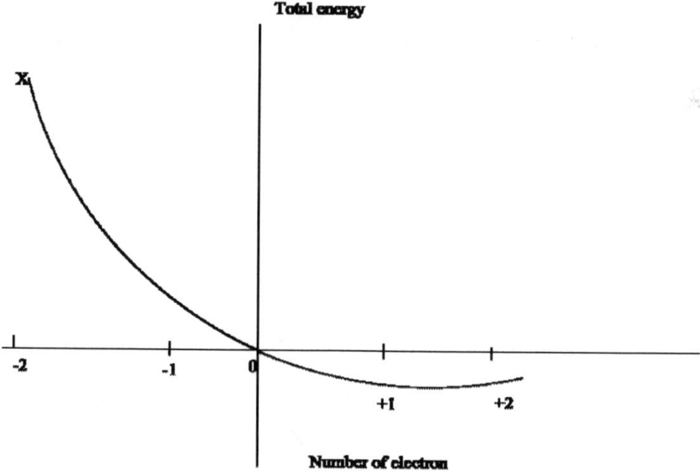

Figure-1: Plot of the energy change with the degree of ionization.

They found that for an atom or ion, the slope at the origin (-dE/dN) , of the E vs. N plot can be identified as electronegativity.

χ= (-dE/dN) (33)

Klopman [55] proposed that '*the atomic terms for any atom can be defined as the sum of those integrals in which the Hamiltonian represents the interaction of the core of the atom with the electron around it*' and critically justified the Iczkowski and Margrave electronegativity concept as under-

The E versus N relationship is usually linear so the higher terms of the equation (32) can be neglected. This leads to the simplified expression (34)

$E = aN + bN^2$ (34)

On differentiation with respect to N, we obtain-

$\partial E/\partial N = a + 2b\,N$ (35)

The electronegativity according to the Pauling definition can be represented [55] by the potential around the atom thus it can be represented by ($\partial E/\partial N$).

$\chi=\partial E/\partial N = a + 2b\,N$ (36)

Iczkowski and Margrave's definition of electronegativity has found widespread acceptance. Hinze, Whitehead and Jaffe [56] opined that electronegativity is not an atomic property, but the property of an orbital of an atom(X or Y) in a molecule(XY) and thus it is dependent on the valence state of the atom(X or Y). Kolpman[55] also opined that electronegativity is an orbital characteristic and therefore the both the ionization potential and electron affinity have to be measured for the same orbital.

Now, for the valence state of an atom, A, N=1,

$$E=a+b=I \qquad (37)$$

where, $a = (3I – A) / 2$ and $b = (A – I)/2$

And for the valence state of an anion, A^-, N=2,

$$E=2a+4b=I+A \qquad (38)$$

Here, 'I' is the ionization energy and 'A' is the electron affinity of the atom (A or B) in its valence state.

When N = 1 the electronegativity leads to an expression similar to that proposed by Mulliken [34].

$$\chi =- (\partial E/\partial N)_{N=1} = a + 2b = (I + A)/2 \qquad (39)$$

Finally, Klopman[55] pointed out that "*however, that in order to represent the electronegativity of an atom in a molecule correctly, the atom must be considered in its valency state and this requires the introduction of electron spin*"

Mulliken's concept of electronegativity explanation does not exclude changes in hybridization. It is transparent from Mulliken concept fact that electronegativity is an orbital characteristic thus the ionization potential and the electron affinity both can be measured for the same orbital.

2.9. Hinze and Jaffe scale for orbital electronegativity of neutral atoms (1962):

Hinze and Jaffe [39] opined that electronegativity is not a property of atoms in their ground state, but of atoms in the same condition in which they are found in molecules, the valance state. They also noticed that the electronegativity can be defined in terms of bonding orbital and the term "Orbital electronegativity" is then suggested.

In the next year Hinze, Whitehead and Jaffe [56] proposed that the orbital electronegativity can be a measure of the power of an atom. It may also be exist in a molecule. The power of an atom to attract an electron in a given orbital to itself can thus be correlated with the orbital electronegativity. The orbital electronegativity is defined as the derivative of the energy of the atom respect to the charge in the orbital, i.e., the number of electrons in the orbitals-

$$\chi_j = \partial E / \partial n_j \qquad (39)$$

Where, χ_j is the orbital electronegativity of the jth orbital and n_j is the occupation number of the jth orbital.

This definition implies two assumptions – (a) that the occupation number n_j may have both integral and non-integral values and, (b) that once assumption (a) is made, than the energy E is a continuous and differentiable function of n_j.

Thus,

$$\chi_j = \partial E/\partial n_j = b + 2c\, n_j \qquad (40)$$

where b and c are the coefficients.

2.10. Yuan scale of electronegativity (1964):

Yuan [57] introduced a scale of electronegativity as the ratio of the number of valence electron to the covalent radius. This scale was later modified by Luo et al [58] on the basis of the number of valance electrons in the bonding atoms and covalent radius of the atom.

2.11. Gyftopoulos and Hatsopoulos quantum thermodynamic scale of electronegativity (1968):

In 1968, Gyftopoulos and Hatsopoulos [59] defined electronegativity on quantum thermodynamical basis. They considered a free atom or an ion as a thermodynamic system. The electrochemical potential measure the escaping tendency of a component from a thermodynamic system, thus 'the power to attract' or the electronegativity is the negative of the escaping tendency or chemical potential.

They defined the electrochemical potential (μ) of a thermodynamic system as

$\mu = (\partial E/\partial N)_{entropy}$ (41)

where, N is the number of electron

The electronegativity of the system was identified as the negative of the chemical potential by them.

$\chi = -\mu$ (42)

2.12. Phillips scale of electronegativity (1968):

Phillips [60] scale of electronegativity is based on the dielectric properties of atom. In addition, Chen [61] proposed an empirical expression of electronegativity using charge-radius ration and some other parameters of atomic structures describing the radial distribution of the inner electron.

2.13. St. John and Bloch quantum defect scale of electronegativity (1974):

St. John and Bloch [62] suggested that some dimensionless parameters which are directly derived from atomic spectral data can be used to define a scale of electronegativity for non transition elements. They invented that the Pauling force potential model, which provides the solution of one electron Schrodinger equation, can be successfully applied in studies of atoms, molecules and solids. The eigen values contain the quantum defect which is physically related to the depth of the potential well and the strength of the effective centrifugal barrier. St. John and Bloch proposed that it particularly convenient to express these quantities in terms of the positions of the radial maxima of the unscreened, lowest valence eigen functions of the Pauling force model potential. For the non transition elements the s-p hybridization can be reflected in a structural index, S, suggested by Bloch and Simons [63]. Now St. John and Bloch redefined S in terms of Orbital electronegativity as-

$$S=(\chi_0-\chi_1)/\chi_0 \qquad (43)$$

where χ_l is the orbital components which measures the scattering power of the core for the lth particle wave.

The sum of the orbital components of the electronegativity is proportional to the total electronegativity of the atom.

$$\chi \propto \sum_{l=0}^{2} \chi_l \qquad (44)$$

They found a linear relationship with the Pauling electronegativity and proposed the quantum defect scale of electronegativity as-

$$\chi = 0.43\sum_{l=0}^{2} \chi_l + 0.24 \qquad (45)$$

They applied their scale for explaining the structures, chemical properties etc for the simple solids such as zinc-blende and wurtzite.

2.14. Density functional scales of Electronegativity (1978):

Electronegativity has been one of the most accepted and used concepts in chemistry for more than 60 years due to the original work by Pauling however, its physical significance has been elucidated in terms of the density functional theory[65] by Parr et al, who, following Iczkowski and Margrave[54], have justified that electronegativity is the negative of the chemical potential of an electronic system, such as an atom or a molecule, and simultaneously demonstrated that electronegativity is constant throughout an atom or a molecule, which validates Sanderson's[47(d)- (f)] postulate that when two or more atoms combine to form a molecule, their electronegativities get equalized. This established a new basis for understanding and application of electronegativity.

2.14.1 Parr electronegativity scale (1978):

Following Iczkowski and Margrave, the definition of electronegativity was introduced by Parr et al [64] on the basis of Density functional theory [65] as-

$$\chi=-\mu=-(\partial E/\partial N)_v \quad (46)$$

where μ is the chemical potential [59] of the system.

2.14.2 Parr and Pearson finite difference scale of electronegativity (1983):

In 1983, Parr and Pearson [66] using finite difference approximation made an attempt to propose an analytical form of electronegativity and hardness. The concept of chemical hardness is very old in chemistry whose basis lies on some experimental observations by various inorganic chemists. Hardness (or inverse of hardness, known as 'softness') is an intrinsic property of atoms and molecules which signify the deformability of atoms and molecules under small perturbation. More discussion on this topic is out of scope of this work. However we proceed to revisit the relationship between DFT, electronegativity and hardness simultaneously in this section.

The chemical hardness, electronegativity and DFT came together in the year of 1983 at the Institute for Theoretical Physics in Santa Barbara. It was a great step forward when Parr showed figure-2 to Pearson where the total electronic energy of a chemical system in its different states of oxidation i.e positive, neutral and negative states is plotted as a function of number of electrons of those systems. Parr asked him whether the curvature of the curve, in the way in which the slope changes with N, i.e $(\delta^2E/\delta N^2)_{V}$, is related to hardness. Pearson applied the Finite Difference (FD) approximation method and found an operational formula for $(\delta^2E/\delta N^2)_{V}$ which was (IP–EA). Pearson exclaimed that it was exactly what he meant by hardness in his landmark 'Hard Soft Acid Base' paper [41]! Then Parr and Pearson, using the density functional theory (DFT) as a basis have rigorously defined the term hardness.

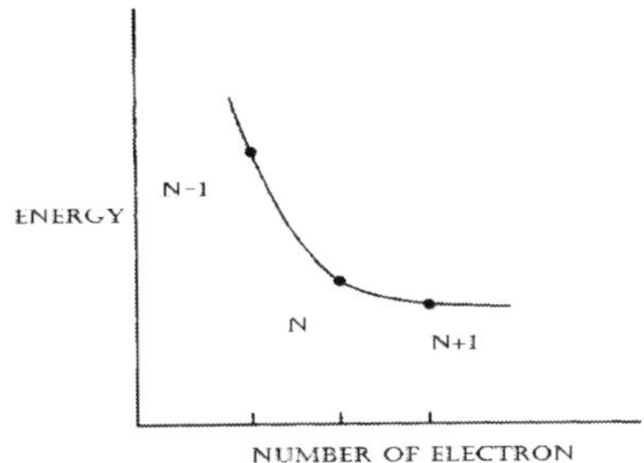

Figure-2: Plot of total electronic energy (all are negative) of a system in positive (+1), neutral (0) and negative (-1) state as a function of number of electrons (N).

Parr et. al. [64] discovered a new fundamental quantity as a new index of chemical reactivity known as the electronic chemical potential (μ). The chemical potential (μ) is a characteristic property of atoms, molecules, ions and radicals and is the first derivative of the energy expression (34).

The second derivative; $(\delta^2E/\delta N^2)_V$ of the E vs N curve is the absolute hardness of the species .i.e.,

$$\eta = 1/2\ \{(\delta^2E/\delta N^2)_V\} \qquad (47)$$

The softness(S) is the reciprocal of the hardness; $S = 1/\eta$ (48)

Quantum mechanics provides us to write the energy of the valence electrons in the form of the following quadratic approximate equation (34) as

$$E = aN + b\,N^2$$

where 'a' is a constant and it is a combination of core integral and a valence shell electron pair repulsion integral and 'b' is half of the average valence shell electron- electron repulsion integral.

Now differentiating equation (34) with respect to N at constant external potential,v, Parr and Pearson proposed-

$-(\partial E/\partial N)v = -a-2bN= (IP+EA)/2= \chi_M$ (49)

χ_M is the Mulliken electronegativity.

Now, second derivative of the above equation (50) leads to-

$(\partial^2 E/\partial N^2)v =(\partial \chi /\partial N)v= b= (IP-EA)/2$ (51)

This expresses the hardness (η) of the corresponding system.

$\eta= b=(IP-EA)/2$ (52)

Using Finite Difference (FD) approximation, the Mulliken electronegativity can be recovered from the Parr electronegativity definition-

$\chi_{Parr}=-\mu= -(\delta E/\delta N)_V$

or, $\chi_{Parr} = - (E_N- E_{N-1})/2 = \{(E_{N-1-} E_N)+(E_N- E_{N+1})\}/2 = (IP+EA)/2= \chi_M$ (53)

This electronegativity scale is known as absolute scale of electronegativity.

2.14.3 Pearson frontier orbital scale of electronegativity (1986):

In 1986, within the limitations of Koopmans' theorem, Pearson [67] putted electronegativity into a MO framework.

The orbital energies of the frontier orbitals are given by

$-\varepsilon_{HOMO} = I$ (54)

and

$-\varepsilon_{LUMO} = A.$ (55)

Thus on the basis of frontal orbital theory, he achieved

$\chi = (\partial E/\partial N)v$

$= (I+A)/2$

$\chi = -(\varepsilon_{LUMO} + \varepsilon_{HOMO})/2$ (56)

and hardness,

$\eta = (\partial^2 E/\partial N^2)v = (I-A)/2$

or, $\eta = (\varepsilon_{LUMO} - \varepsilon_{HOMO})/2$ (57)

He again pointed out that a hard species has a large HOMO-LUMO gap and a soft species has a small HOMO-LUMO gap(Figure-3).

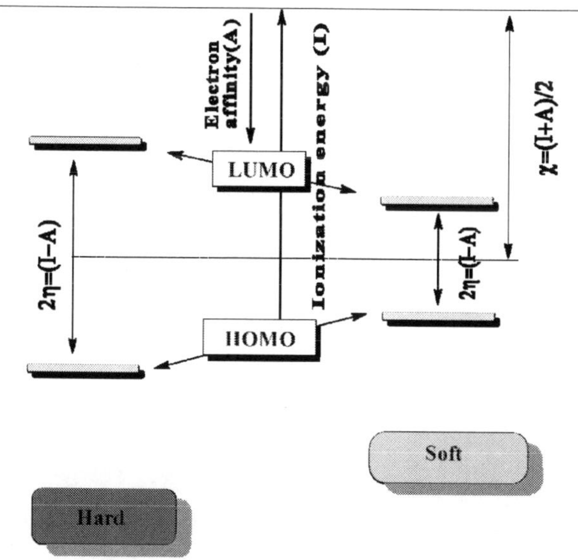

Figure-3: HOMO-LUMO gap

2.15. Pasternak scale of electronegativity (1978):

Pasternak [42] considered the property of electronegativity in simple bond charge (SBC) model of diatomic molecule and proposed that electronegativity of the atom X and Y in the XY molecule is proportional to the ratio of the nuclear charge of the atom X and Y and ½ of the bond length in XX and YY respectively.

$$\chi_X=C(Z_X/r_X) \quad (58)$$

and,

$$\chi_Y=C(Z_Y/r_Y) \quad (59)$$

where C is a constant depends on bond type.

2.16. Zhang scale of electronegativity (1982):

Zhang's [68] scale is based on the ionization energies and the covalent radius of the atom. Zhang defined the term 'electronegativity' as the electrostatic force (F) excreted by the effective nuclear charge (Z_{eff}) on the valence electrons.

i.e, $F \propto Z_{eff}/r_{cov}$ (60)

Now the Slater definition of I is-

$I=RZ^2_{eff}/n^{*2}$, (61)

Substituting Z_{eff} in equation (60) he achieved

$F \propto n^*(I/R)^{1/2}/r^2_{cov}$ (62)

Scaling with Pauling value, he proposed the electronegativity ansatz as

$\chi=0.241\{n^*(I/R)^{1/2}/r^2_{cov}\}+0.775$ (63)

2.17. Boyd and Edgecombe scale of electronegativity (1988):

Boyd and Edgecombe [69] proposed an atomic electronegativity scale based on the topological properties of the electron density distributions of molecules, and they extended this method to evaluate group electronegativities.

In this work, Boyd and Edgecombe assumed that there is an electronegativity factor (F_A) associated with atom A that is directly proportional to The distance of the bond critical point from the hydrogen atom in AH (r_H) and inversely proportional to the electron density at the bond critical point, $p(r_c)$, where r_c

denotes the position of the bond critical point. They also observed that this factor fails to allow for the larger size of the heavier atoms.

They define a term 'orbital multiplier' f_{AB}, as

$f_{AB} = R_A/(R_A + R_B)$ (64)

where R_A and R_B are the distances from the nuclei to the orbital center.

They stated that the deviation of f_{AB} from 0.5 measures the difference in the electron-attracting power, or electronegativity, of atoms A and B.

They pointed out that $p(r_c)$ increases monotonically within each period as the atomic number of A, Z_A, increases. Thus, $p(r_c)$ increases while r_H decreases. Boyd and Edgecombe assumed that there is an electronegativity factor (F_A) associated with atom A that is directly proportional to r_H and inversely proportional to $p(r_c)$. They also assumed that as the electronegativity increases from left to right within each period while r_H decreases, the electronegativity factor varies inversely with the number of valence electrons of the atom A, N_A. Boyd and Edgecombe also stated that the electronegativity factor concept fails to allow for the larger size of the heavier atoms.

They define the term electronegativity factor as

$F_A = r_H/N_A \, p(r_c) r_{AH}$ (65)

They expressed the electronegativity of atom A as a power curve of F_A

$\chi_A = aF^b$ (66)

They sated the value of the two parameters as a =1.938 and b=-0.2502 to provide the electronegativities of Li and F as1 .00 and 4.00 respectively and evaluated the atomic electronegativity of the 21 elements of second, third and fourth period.

2.18. Allen scale of electronegativity (1989):

Perhaps the simplest definition of electronegativity is that of Allen [70], who stated that electronegativity is related to the average energy of the valence electrons in a free atom.

He proposed the electronegativity ansatz -

$\chi_{\text{Allen}} = (n_s\varepsilon_s + n_p\varepsilon_p)/(n_s + n_p)$ (66)

where ε_s and ε_p are the one-electron energies of s- and p-electrons in the free atom and n_s and ,n_p are the number of s- and p-electrons in the valence shell respectively. It is usual to apply a scaling factor, 1.75×10^{-3} for energies expressed in kilojoules per mole or 0.169 for energies measured in electron volts, to give values which are numerically similar to Pauling electronegativities.

Furthermore, Allen considered electronegativity as configuration energy of the atoms of interest [13, 71] and he stated that "*When orbital occupancy is taken into account, it immediately follows that configuration energy (CE), the average one-electron valence shell energy of a ground-state free atom, is the missing third dimension*".

He suggested the ansatz (67) and (68) for the calculation of the configuration energies (CE) or the electronegativities of the s-p and d-block elements.

For s-p block elements

$\chi_{s\text{-}p}=(CE)_{s\text{-}p}=(n_s\varepsilon_s+ n_p\varepsilon_p)/(n_{s+}n_p)$ (67)

And for the atoms with ground-state configurations $s^n d^m$ and $s^{n-1}d^{m+1}$

$\chi_d=(CE)_d =(p\varepsilon_s+ q\varepsilon_d)/(p+q)$ (68)

where ε_s and ε_d are the multipulate-averaged one-electron energies of s- and d-orbitals of the atom in the lowest energy configuration respectively. In the free atom n and m are the usual integers such that (p + q) is the maximum oxidation state observed for the atom in any compound or complex ion.

The multipulate-averaged one-electron energies can be directly determined from spectroscopic data, and so the electronegativities calculated by this method are originally referred to as spectroscopic electronegativities by Allen. The necessary data are available for almost all elements, and this method allows the estimation of electronegativities for elements which cannot be treated by the other methods. However, for d- and f-block elements, doubt in the electronic configuration may arise for the calculation of the electronegativity by Allen method.

2.19. Nagle scale of electronegativity (1990):

Nagle [72] scale of electronegativity is based on atomic polarizability. The static electric dipole polarizability or simply polarizability is an experimentally measurable property of an isolated atom. The valence electron density is a parameter which can define and measure the electronegativity of an atom. Nagle found that the cube root of this ratio of the number of valence electrons divided by the polarizability, $(n/\alpha)^{1/3}$, can be used as a measure of electronegativity for all s- and p-block elements (except the noble gases). The value fits well with the electronegativities in Pauling scale and the correlation yields-

$\chi = 1.66 (n /\alpha)^{1/3} /+ 0.37$ (69)

2.20. Zheng and Li scale of electronegativity (1990):

Zheng and Li [73] scale is base on the average nuclear potential of the valence electron. Zheng and Li discovered a new method for the determination of the effective nuclear charge Z_{eff} of the atoms. They proposed a new electronegativity scale equivalent to the Mulliken electronegativity scale by introducing the valence electron's mean radius $<r>_{nl}$ in to the Mulliken electronegativity (I+A)/2 and defined electronegativity as the ratio of Z_{eff} and $<r>_{nl}$.

i.e., $\chi_M=(I+A)/2= Z_{eff}/ <r>_{nl}$ (70)

2.21. Ghosh scale of electronegativity (2005):

Ghosh electronegativity scale [74] is based on the absolute radii of atoms. In this study, Ghosh pointed out that the electronegativity is an intrinsic free-atom periodic property like the atomic size. Electronegativity and atomic radius both are periodic property and inversely related through the periodic table. He put forward an ansatz for the evaluation of electronegativity in terms of the absolute radii of atoms as-

$$\chi = a\,(1/R) + b \qquad (71)$$

where χ is electronegativity and R is absolute radius of atoms, a and b are two constants determined by least square fitting for each period of elements separately.

2.22. Ghosh and Gupta electronegativity scale (2006):

Ghosh and Gupta [75] electronegativity scale is based on the polarizability *(α)* of atoms. They suggested that both polarizability and electronegativity are periodic properties and they are connected by an inverse relationship. Relying upon their express behavior along the periods of the periodic table, a general and simple relation between χ and α is suggested by Ghosh and Gupta as-

$$\chi = m(1/\alpha)^{1/3} + c \qquad (72)$$

where m and c are constants are evaluated by least square fitting by plotting χ vs.$(1/\alpha)^{1/3}$ for each period separately. Ghosh and Gupta evaluated the electronegativity of 54 elements of periodic table using the ansatz (72).

2.23. Keyan Li and Dongfeng Xue ionic electronegativity scale (2006):

On the basis of absolute electronegativity theory of Parr and Pearson, Keyan Li and Dongfeng Xue [76] proposed an electronegativity scale for the elements in different valence states and with the most common coordination number in terms of effective ionic potential.

They defined the electronegativity of an element as “the electrostatic potential at the boundary of an ion caused by its effective nuclear charge”.

The effective ionic potential was defined by them as

$$\varphi=n^*(I/R)^{1/2}/r \qquad (73)$$

where $I_m\{ I_m =R(Z_{eff}/n^*)\}$ is the ultimate ionization energy, n^* is the effective principal quantum number and R(R=13.6 eV)is the Rydberg constant and r_i is the ionic radius.

They considered electronegativity of an ion is proportional to the effective ionic potential and proposed a scale for ions through a linear regression with Pauling electronegativity as follows-

$$\chi_{ion}=0.105\ \varphi+0.863 \qquad (74)$$

They calculated the electronegativities of 82 elements in different valence states and with the most common coordination numbers using the above ansatz and found that for a given cation, the electronegativity increases with increasing oxidation state and decreases with increasing coordination number.

Some important chemical phenomena, such as the ligand field stabilization, the first filling of p orbitals, the transition-metal contraction, and especially the lanthanide contraction, are well-reflected in the relative values of the proposed scale of electronegativity by Keyan Li and Dongfeng Xue. The scale can also be use to quantitatively estimate the Lewis acid strength for the main group elements in their highest oxidation state.

2.24. Noorizadeh and Shakerzadeh scale of electronegativity (2008):

Noorizadeh and Shakerzadeh [77] proposed a new electronegativity scale based on electrophilicity index. In this paper, they pointed out that the electrophilicity ($\mu^2/2\eta$ or $\chi^2/2\eta$) of a system is related to both the resistance and the tendency of the system to exchange electron with the environment. Thus the electrophilicity index can be used to measure the electronegativity of the system. To calculate the electrophilicity index of a system, they use the B3LYP/6-311++G level of theory and evaluated the energy of atom in 0, -1,-2,+1 and +2 states. They take the energy to be a Morse like function of the number of electron and they correlate the evaluate electrophilicity index with Pauling and Allred and Rochow scale.

2.25. Ghosh and Islam scale of electronegativity (2009):

Ghosh and Islam [10] recently pointed out the conceptual commonality between the two fundamental theoretical descriptors, electronegativity and hardness. They concluded that the hardness and the electronegativity originate from the same source, the electron attracting power of the screened nucleus upon valence electrons and discovered the surprising result that if one measures hardness, the electronegativity is simultaneously measured and vice-versa.

They proposed probably the simplest electronegativity ansatz as

$\chi=\eta$ (75)

3. Common proposition regarding electronegativity:

A search of literature [13] reveals that a good number of workers converge to number of common proposition regarding electronegativity:

1. Electronegativity is a periodic property.
2. Electronegativity is an intrinsic atomic property which is associated with shell structure of atoms and arises from the screened nuclear charge.
3. It is a global property of atoms, molecules and ions.
4. It is a property which has to be measure in energy units.

Although 70 years have over and done since the electronegativity concept was given quantitative expression, the attempts to refine electronegativity theory are not yet reached to the final scale of electronegativity.

Ghosh and Islam [10] recently mentioned that the as electronegativity is a conceptual entity, not an experimentally observable property and as there is no quantum mechanical operator for electronegativity can be suggested; it is empirical and will empirical.

So we will continue to make use of the "crude" electronegativity scales for the correlation of a vast field of chemical knowledge and experience.

There are certain rules for a reasonable scale of electronegativity. Scientists [8, 13, 78, 79] laid down certain sine qua non of certain electronegativity scale, viz,

1. The scale has a free atom definition.
2. Electronegativity should be expressed in energy unit.
3. Contraction of the main transition group elements must be transparent.
4. Electronegativity value of noble gas elements is very high.
5. The scale must satisfy the Silicon rule: All metals must have Electronegativity values that are less than or equal to that of Si.

6. Carbon rule: The Electronegativity value of C has to be greater than, or at least equal to, that of H.

7. Metalloid band: the six metalloid elements B, Si, Ge, As, Sb and Te that separate from the non metals have electronegativity values, which do not allow overlaps between metals and non metals.

8. The scale should quantify the Van Arkel-Ketelaar triangle.

9. A high precision is necessary for each scale.

10. In binary compounds, the electronegativity of the constituent atoms clearly quantifies the nature of bonds.

11. χ must have a quantum mechanically viable definition. That is, it must be compatible with the elementary quantum concepts such as shell structure, quantum numbers, and energy levels which describe the electronic structure of atoms.

4. Unit of electronegativity:

We know that it is very difficult to understand the meaning of a quantity if one does not know its unit properly. The physical picture corresponds to the term 'electronegativity' is till now not clear to us. Each scale has its own identity and usefulness in the field of application. They are not comparable to each other, thus the units of different scales are different. We have quoted here the units a few scales below.

Scale	Dimension
Pauling[6]	$(\text{Energy})^{1/2}$
Mulliken [34], Ghosh[74], Ghosh Gupta[75], Ghosh Islam[10]	Energy
Allred and Rochow[7]	Force
Gordy[38]	Energy / electron
Sanderson[47] ,Gordy[17]	Dimensionless
Parr[64]	Energy / electron
Allen[13, 70, 71]	Average one- electron energy
Walsh[46]	Force/distance

Table-1: Some popular scales of electronegativity and their dimension

5. Electronegativity and other Periodic parameters:

It is now well established fact that, like ionization potential, atomic radius, and others electronegativity is also a periodic parameter [80, 45, 53, 74-75].

When we look in the total periodic table we are convinced that in a period electronegativity would increase monotonically to be maximum at the noble gas elements and in the pattern is repeated next period and for each row, electronegativity would decrease monotonically to be minimum at the last element of each row. The periodic law is a very fundamental law of nature manifested in many physico-chemical properties of atoms and molecules. Although the periodic table does not follow the quantum mechanics, it has chemical organizing power relating many seemingly different properties which are individually periodic. So, one periodic parameter can be converted to another. We already discuss in section I, that various attempts have been made to evaluate electronegativity of atoms using other periodic parameters, like atomic radius[47,74], ionization potential[68], polarizability[72,75], hardness etc. but there are certain works which actually does not suggest a scale of electronegativity but suggested a relationship between electronegativity and other periodic parameters. Let us have a look on few of them-

Pearson [81] suggested that for donor atoms, the electronegativity can be taken as a measure of the hardness of the base. After rigorous research on systematic formulation of electronegativity and hardness, Putz [82] opined that the hardness and electronegativity are proportional to each other-

$$\chi \propto \eta \qquad (76)$$

Ayers [83] on the basis of the energy expression of March and White [84] proposed expressions for the electronegativity and the hardness of neutral atoms and pointed out that the two fundamental atomic parameters; hardness and electronegativity are proportional to each other.

Conclusion:

From the above discussions, it is self evident that no rigorous definition of electronegativity has been suggested and the final scale of electronegativity is yet to develop. Ghosh and Islam[10] opined that electronegativity is not an observable property and hence, no quantum mechanical operator can be assembled for its quantum mechanical evaluation. It is an empirical quantity and remains empirical. So there is a plenty scope of research on this topic and actually it is now an animated field of current research. Allen [8, 70-72] suggested that the concept and scale of electronegativity have a 'broken symmetry' symmetry relationship with Periodic Tables categorization, which completes the Periodic Table. Following Pauling, some scientist believed that electronegativity is an in situ property developed on molecule formation rather it is an intrinsic ground state property of atom and it is carried in to molecules but a group of scientist, such as Allen[8,70-72], Murphy et al[78], Parr et al[64] etc have conclusively established that electronegativity is an atomic property. Majority of workers recommend that electronegativity is not an in situ property developed on molecule formation rather it is an intrinsic ground state property of atom and it is carried in to molecules. Allen and Knight [86] opined that the in situ assumption is self defeating and so the electronegativity is very difficult to define electronegativity.

The concept of electronegativity and electron attracting power of an atom bonded to divergent atoms are now accepted as true *'in each other's pocket'*. This electron attracting power begins from the effective nuclear charge. It, therefore, transpires that electronegativity is a fundamental property of atomic shell structure and obviously periodic in nature.

From this discussion, we may conclude that the electronegativity is a fundamental descriptor of atoms molecules and ions which can be used in correlating a vast field of chemical knowledge and experience. The attempts to refine the concept and scale of electronegativity theory are not yet sufficiently complete to enable a judgment to be reached on their effectiveness. Thus there is a plenty scope of research on this topic for the further development. We quote original from Pritchard and Skinner, "*Meanwhile, it seems safe to say that the chemist will continue to make use of the crude electronegativity theory for some time yet-a practice for which he can hardly be blamed in the absence of an alternative theory of equal generality*".

Reference:

[1] A. Avogadro, J. de. Phys. 73, (1811), 58.

[2] A. Volta, Philos Mag, Sept, (1800).

[3] C.Heinrich Pfaff,; H. H. U Vitalis Pfaff, Neuere Geometrie, Scholarly Publishing Office, University of Michigan Library.

[4] The Pauling Electronegativity Scale: Part 1, Historical Background, paulingblog.wordpress.com.

[5] J.J. Berzelius, (a) Annals. Philo. **2**, (1813),443. (b) ibid,, **3**,(1814), 51, 93, 244, 353.

[6] (a) L. Pauling, The Nature of the Chemical Bond.3rd edn.;Cornell University:Ithaca.NY .(1960), (b) ibid, J. Am. Chem. Soc., ,54, (1932), 3570. (c) Pauling, L, Yost, D.M., Proc. Nat. Acad. Sci., 18, (1932),414.

[7] Allred, A. L.; Rochow, E. G., J. Inorg. Nucl. Chem. 5, (1958), 264.

[8] Allen, L.J. Am. Chem. Soc. 114, (1992), 1510.

[9] Huheey, J. E. Inorganic Chemistry; Harper and Row New York, 1975.

[10] D.C. Ghosh, N. Islam, Int. J. Quantum Chem., 109, (2009),in press.

[11] K. Fukui, Science. 218, (1982),747.

[12] H.O Pritchard.; Skinner, H.A. Chem. Rev. 55, (1955), 745.

[13] D.C Ghosh, J. Indian Chem. Soc, 80,(2003), 527.

[14] K.D.Sen,; C.K. Jorgensen, Electronegativity, Springer-Verlag, New York, 1987.

[15] R.R. Reddy.; T.V.R Rao,; R. Viswanath, J. Am. Chem.Soc. 111, (1989),2914.

[16] R. T. Myers,. J. Chem. Educ. 56, (1979), 711.

[17] W. Grody, J. Chem. Phys., 19, (1951),792.

[18] H.S. Gutowsky, C. J. Hoffman J. Chem. Phys.,19, (1951)1259.

[19] K.S. Lackner; G. Zweig, Phys.Rev.D.(1983), 28, 1671.

[20] U. Wahl, E.Rita,; J. G Correia,.; E. Marques, A. C. Alves,; J. C Soares,. Phys. Rev. Lett. 95, (2005), 215503.

[21] R. Asokamani, R. Manjula,., Phys. Rev. B.,39, (1989), 4217.

[22] S.G Hur,T.W. Kim,.S. HungJ. Phys. Chem. B., 109, (2005),15001.

[23] L. Zhang; ,E. G Wang.; Q. K. Xue, S. B Zhang,.; Z. Zhang. Phys. Rev. Lett., 97, (2006), 126103.

[24] M. E Arroyo-De Dompablo.; M. Armand; J. M. Tarascon; U Amador. ; Electrochemistry communications 8, (2006),1292.

[25] S. Kobayashi, H. Hamashima,; N Kurihara,Miyata,; A. Tanaka, Chem. Pharm. Bull., 46, (1998),1108,.

[26] E.M Zueva, V.I. Galkin, A.R Cherkasov,; R.A. Cherkasov,. Russ. J. Org. Chem. 38, (2002), 613.

[27] M. Haissinsky,. J. Phys. Radium. 7, (1946), 7.

[28] M.L Huggins. J. Amer. Chem. Soc.,75, (1953), 4123.

[29] W. Gordy, W.J. Orville Thomas, J. Chem. Phys, 24, (1956), 439.

[30] A.P Altshuller,. J. Chem. Phys,22, (1954), 765

[31] A.L Allred,. J. Inorg. Nucl. Chem., 17, (1961), 215.

[32] R. S Mulliken,. J.Chem. Phys. 3, (1935), 573.

[33] J.G. Malone, J. Chem.Phys, 1, (1933), 197,

[34] R. S. Mulliken, J. Chem. Phys., 2, (1934), 782.

[35] F.Hund, Zeits. f. Physik,73, (1931),1.

[36] C.A. Coulson, Proc R Soc London Ser A. 207, (1951), 63.

[37] R.G. Parr, R.G. Pearson, J. Am. Chem. Soc, 105, (1983), 7512.

[38] W. Gordy, Phys.Rev. 69, (1946), 604.

[39] J.Hinze, H.H. Jaffe, J.Am.Chem. Soc., 84, (1962),540.

[40] JE. Huheey, (a)J. Phys. Chem. 69, (1965), 3284; (b) ibid ,70, (1966),2086.

[41] R.G.Pearson, J. Am. Chem. Soc. 85, (1963), 3533.

[42] A.Pasternak, Chem.Phys., 26, (1977), 101.

[43] N.K.Ray, L. Samuels, R.G.Parr, J. Chem. Phys., 70, (1979,)3680.

[44] P.Politzer,; R.G. Parr, D.R.Murphy, J.Chem.Phys.79, (1983),3859.

[45] D.C.Ghosh, , T. Chakraborty, J. Mol. Struct: THEOCHEM 906, 2009,87.

[46] A.D. Walsh, Proc. Roy. Soc.(London) A, 207, 1951,13.

[47] R. T.Sanderson (a) J. Chem. Edu. 29, (1952),539; (b) ibid, 65, (1988),112. (c) ibid, J. Am. Chem. Soc.105, (1983), 2259. (d)ibid, Science, 114 , (1951), 670; (e) ibid, ,116, (1952), 41; (f) ibid, 121, (1955), 207.

[48] E.J Little, M.M Jones. J Chem Educ,37, (1960),231.

[49] C.Mande, P. Deshmukh, P. Deshmukh,. J. Phys. B: At. Mol. Phys. 10, (1977),2293.

[50] R.J. Boyd, G.E. Markus, J Chem Phys, 75, (1981),5385.

[51] R.J.Boyd, J Phys B, 10, (1977), 2283.

[52] E. Clementi, C. Roetti., At. Data Nucl Data Tables, 14, (1974),177.

[53] D.C.Ghosh, T.Chakraborty, B. Mandal, Allred Rochow's Scale of Electronegativity Revisited, Theor. Chem. Account. 2009,in press.

[54] R.P.Iczkowski, J. L. Margrave, J. Am. Chem. Soc. 83, (1961), 3547.

[55] G.Klopman, J.Am.Chem.Soc.,86, (1964), 4550.

[56] J. Hinze, M.A.Whitehead, Jaffe. H.H. J. Am. Chem. Soc, 85, (1963),148.

[57] H Yuan,.J, Acta Chimica Sinica, 30, (1964),341.

[58] Luo Y.R., S. W.Benson, (a) J. Phys. Chem. 92, (1988), 5255; (b) ibid, J.Am. Chem. Soc. 111, (1988), 5255.(c) Luo, Y.R. ; Pacey, P. D. J. Am. Chem. Soc., 113, (1991), 1465.

[59] E.P. Gyftopoulos, G.N. Hatsopoulos, Proc. Natl. Acad. Sci. US, 60, (1965),786.

[60] J.C.Phillips, (a) Phys Rev Lett 20, (1968),550 (b)ibid, Covalent Bonding in Crystals, Molecules, and Polymers, University of Chicago Press, Chicago, IL, 1969.

[61] Chen N.Y., Chang, H.K, Acta. Chimica. Sinica, 33, (1975),101.

[62] St. A.John, N. Bloch, Phys. Rev. Lett. 33, (1974),1095.

[63] A.N. Bloch, G. Simons, J.Am Chem. Soc, 94, (1972),8611.

[64] R. G.Parr, R. A. Donnelly, M.Levy, W. E. Palke, J. Chem. Phys, 68, (1978), 3801.

[65] (a)R.G. Parr , W.Yang, Density Functional Theory of Atoms and Molecules, Oxford University Press, 1989. (b) P.Hohenberg , W.Kohn. Phys. Rev. B, 136, (1964), 864.

[66] R.G Parr. R.G. Pearson, J. Am. Chem. Soc. 105, (1983), 7512.

[67] R. G. Pearson, Proc. Natl. Acad. Sci. (1986), 83, 8440.

[68] Y. H Zhang,. Inorg. Chem., 21, (1982), 3886.

[69] R. J Boyd, K.E. Edgecombe, J. Am. Chem. Soc.110, (1988), 4182.

[70] L. C. Allen, J. Am. Chem. Soc. 111, (1989), 9003.

[71] L. C. Allen, J. Am. Chem. Soc. 122, (2000), 5132.

[72] J.K. Nagle, J. Am. Chem. Soc., 112, (1990), 4741.

[73] Zheng Neng-Wu, Li Geo-Sheng, J Phys Chem, 98, (1994),3964.

[74] D.C. Ghosh J. Theoret. Comput. Chem., 4, (2005), 21.

[75] D.C. Ghosh, ; K.Gupta, , J. Theor Comput Chem, 5, (2006), 895.

[76] Keyan Li, Dongfeng Xue J Phys Chem. A 110, (2006),11332.

[77] S. Noorizadeh, E. Shakerzadeh, J. Phys. Chem. A, 112, (2008) 3486.

[78] L.R.Murphy, T. L. Meek, A. L.Allred, L.C. Allen. J Phys Chem A. 104, (2000),5867.

[79] M.V. Putz,. Int. J. Quantum. Chem. 106, (2006), 361.

[80] (a) P.K.Chattaraj, B. Maity, J. Chem. Educ., 78, (2001),811, 2. (b) P. K.Chattaraj, D. R. Roy, S. Giri, Computing Letters (CoLe), 3, (2007),223.

[81] R.G. Pearson, Chem. Commun. (1968), 65.

[82] M.V. Putz, Absolute and Chemical Electronegativity and Hardness, Nova Science Publishers, Inc., New York, 2008.

[83] P.W. Ayers, Faraday Discuss,135, (2007),161.

[84] N.H. March, R.J.White, J Phys B (1972), 5.

[85] L.C. Allen, E. T.Knight, J. Mol. Struct. (Theo chem.), 261, (1992) , 313.

[86] L.C. Allen, J. Huheey, J. Inorg. Nucl. Chem., 42, (1980), 1523.

CHAPTER-2:

Application of electronegativity

Index

1.Introduction

2.1. Electronegativity Equalization Principle

2.2. Percentage ionic character

2.3 Hetaropolar bond length

2.4. Electronegativity and molecular orbital theory

2.5.Dipole moment

2.6.Atomic Polar Tensor

2.7. Bond stretching frequency and Force constant

2.8. Standard enthalpies of formation and bond dissociation energy

2.9. Stability ratio

2.10. Lewis acid strength

1. Introduction:

The improvement of structural thinking in chemistry is often based on empirical relations between some properties, which can be measured or can be found in literature and those which one wants to investigate. This is thoroughly useful because both quantities result from the same electronic structure of the molecule under consideration. The ultimate aim of chemists, till now, is to solve the fundamental and naturally occurring phenomena. For this purpose, chemists, rely upon the electronic structure of the fundamental structural construct of the universe-the atom, introduced some very important structural principles in chemistry based on the natural occurring phenomenon and also experimental observations. The concepts which do not originate from a "strong" theory but clearly and vividly describe a series of relations among chemical data are called the empirical entities. They are very important and because of the simplicity they accompany our structural thinking and the technical terms in scientific language. Sometimes they can be improved on the bases of some theoretical methods. As a result, time to time, new concepts have been introduced in chemistry for the rationalization and prediction of various physico-chemical phenomena. The theory approaches a chemical experiment via selective approximations and simplifications which then serve as bridge between the rigorous theory and chemical reality. In

1939, in the first edition of The Nature of the Chemical Bond, Pauling [1] introduced an important fundamental descriptor of not only chemistry but also of science– the electronegativity. Pauling defined electronegativity as “the power of an atom in a molecule to attract electrons to it.” Till then, the electronegativity was defined by Pauling, attempts are made to define and quantify it, but the basic concept of Pauling– “the power of atoms to attract or retain or not to release electrons” is not yet change. There is a maxim that when there are many treatments for a disease, none of them is completely adequate. The same idea could be applied to electronegativity in view of the many attempts to define and quantify it. Allen [2] was the first who recognize that the electronegativity concept and its scale can indeed furnish the appropriate function to reproduce the observed periodicities of experimental quantities.

In the Part-1, we [3] have analyzed the fact that exact definition and the best method for the evaluation of electronegativity remains to be discovered. There is plenty of room for controversy, and no two workers will agree completely. Electronegativity is not an observable property and hence, no quantum mechanical operator can be constructed for its quantum mechanical evaluation. Furthermore, there is no experimental benchmark for the electronegativity In spite of these controversies and the empirical nature, the electronegativity is one of the most useful theoretical descriptor which is widely used by chemists, physicists, biologists and geologists. Some very fundamental conceptual descriptors of organic, inorganic and physical chemistry such as bond energies, bond polarities and the dipole moments, force constants and the inductive effects can only conceived in terms of the concept of electronegativity[4,5]. The significance of the electronegativity concept in chemistry become transparent from the remarks of Coulson [6] "*the astonishing success which the theory has had in correlating a vast field of chemical knowledge and experience*".

Ghosh and Islam[7] hold the idea that the appearance and significance of heuristically developed concepts of electronegativity and hardness in chemistry and physics resemble the unicorns of mythical saga. They exist but never seen. Without the concept and operational significance of hardness and electronegativity, chemistry and many aspects of condensed matter physics becomes chaotic and the long established unique order in chemico-physical world will be disturbed.

The thousand fold application of the electronegativity concept is impossible to review in one work, but we pinpoint some of the "*astonishing successes*" of the electronegativity comcept in the real world and its' relationship with other periodic parameters.

2.1. Electronegativity Equalization Principle:

Understanding chemistry requires a basic understanding of the structure and interactions of atoms, ions, and molecules. Focus on the chemical bond that hold together the atoms that form the molecules, an answer of the fundamental question, why do atoms interact to form molecule was given by the electronegativity equalization principle. After the announcement of the very fundamental law of nature– the electronegativity equalization principle by Sanderson [8], it becomes one of the most useful applications of the electronegativity. To formulate the electronegativity equalization principle Sanderson [8] stated that when two atoms having different electronegativity come together to form a molecule, the electronegativities of the constituent atoms become equal, yielding the molecular, equalized electronegativity. Thus for the first time the concept of electronegativity had been thought of as a dynamic property rather than a static one.

The driving force for the electronegativity equalization can be pictured as follows: Electrons in a stable homonuclear covalent bond are equally attracted to both nuclei. But this is not true in case of a heteronuclear system, where two atoms (or more) having different electronegativity values are joined through covalent bond. The more electronegative atom having more electron attracting power attracts the bonding electron pair more towards itself. Thus some amount of charge transferred from the lower electronegative atom to the higher one. This can be also viewed as charge is transferred from the atom having higher chemical potential value to the atom having lower chemical potential value until both the chemical potential and electronegativity of the constituents becomes equal.

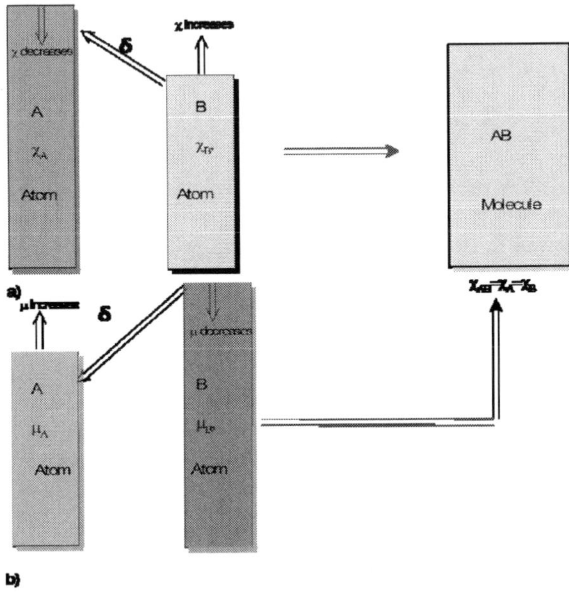

where $\chi_A > \chi_B$

a) Electronegativity equalization

b) Chemical potential equalization

Figure-1: Electronegativity (-chemical potential) equalization scheme.

Two different electronegative atoms have atomic orbitals of different energies. The process of bond formation must provide a pathway by which the energies of the bonding orbitals become equalized. If in the bond formation process the electronegativity of the higher electronegative atom decrease as that atom acquires electronic charge(δ) and that of lower electronegative atom increase as it loses the electronic charge(δ).

Now, how can we calculate the amount of charge transfer and also the intermediate electronegativity in the molecule?

Sanderson [8] postulated a geometric mean principle for the electronegativity equalization. He pointed out that a molecule's final electronegativity is the geometric mean of the original atomic electronegativities. The electronegativity equalization principle is now linked to the fundamental quantum mechanical variation principle. Parr, Donnelly, Levy and Palke [9] identified electronegativity as the amount of energy required to remove a small amount of electron density from the molecule at the point r, i.e.,

$$\chi(r)= \delta Ev(\rho)/\delta\rho(r) \qquad (1)$$

They [9] have noted that the energy is minimized only if the electronegativity is equalized, because if there are two place in the molecule with different electronegativity, then transferring a small amount of electron density, q, from the place to lower electronegativity ($r_<$) to the place with greater electronegativity($r_>$) will lower the energy. Parr and Bartolotti [10] gave a proof of the electronegativity equalization principle from a sound density functional theoretical[11-12] background. The term chemical potential as it occurs in thermodynamics[13] has long been accepted as a perspicuous description of the escaping tendency of a component from a phase. Parr et al [9] identified electronegativity as the negative of the chemical potential of the system. They also pointed out that both parameters can be adopted at the molecular level because they have the very same properties in the charge equalization procedure. Thus they suggested that both the words, "electronegativity" and "chemical potential", can be applied for the electronegativity equalization procedure but they prefer the latter for their discussion.

They correlated Charge Transfer, Electronegativity Difference, and Energy Effect of Charge Transfer with the geometric mean principle [8] of electronegativity equalization.

If a molecule AB in its ground state is formed from atoms A and B in their ground states having chemical potentials μ^0_A and μ^0_B, electron densities ρ^o_A and ρ^o_B numbers of electrons N^0_A and N^0_B, and nuclear potentials υ^0_A and υ^0_B respectively. The molecule formed have the chemical potentials μ_{AB}, electron densities ρ_{AB} and ρ^o_B numbers of electrons $N^0_A + N^0_B = N$ and nuclear potentials $\upsilon^0_A + \upsilon^0_B$.

The the number of electrons which flow from B to A on AB formation was given by Parr and Bartolotti [10] as

$$\Delta N = (1/2\gamma) \ln (\mu^0_B / \mu^0_A) \qquad (2)$$

The value of the γ is not always constant rather it changes in a fairly narrow range.

Parr and Bartolotti[10] estimated the value of the constant $\gamma = 2.15 \pm 0.59$.

The energy difference ΔE is correlated with the standard chemical potential difference of A and B(μ^0_A- μ^0_B) and number of electron transfer ΔN as

$$\Delta E = (\mu^0_A - \mu^0_B)\,\Delta N \qquad (3)$$

One can easily calculate the molecular equalize electronegativity using the Simple Bond Charge(SBC) model[14]. Ray et al[15]first derived the algorithm for the equalized molecular electronegativity and other descriptors such as bond distance, force constants etc using the SBC model [14] and Pasternak[16] electronegativity expression.

For a diatomic molecule AB with the equilibrium bond length R_{AB}, if we consider Z_A and Z_B as the charge on atom A and B in the diatomic molecule AA having the bond length $2r_A$ and the charge on B in BB having the bond length $2r_B$ respectively and δ is the amount of charge transferred during the process of the molecule formation, then $Z_A + \delta$ and $Z_B - \delta$ is the charges on nuclei A and B in the molecule AB.

Pasternak[16] defined the electronegativity of a bonded atom A in a molecule as

$$\chi_A = C(Z_A/r_A), \qquad (4)$$

where C is the constant which depends on type of bond between A and B.

Thus, for the diatomic AB the natural definition for the electronegativity of atom A and B, in the final molecule after charge transfer, has the form-

$\chi_A(\text{in AB}) = C(Z_A + \delta) / r_1,$ (5)

And

$\chi_B(\text{in AB}) = C(Z_B - \delta)/r_2,$ (6)

According to Sanderson electronegativity equalization principle [8], electronegativity of bonded atoms in a molecule must be equal to each other.

Thus,

$\chi_{AB} = \chi_A(\text{in AB}) = \chi_B(\text{in AB})$ (7)

$= C(Z_A + \delta)/r_1 = C(Z_B - \delta)/r_2 = C(Z_A + Z_B)/R_{AB}$

or, $\chi_{AB} = (\chi_A r_A + \chi_B r_B)/(r_1 + r_2) = (\chi_A R_{AA} + \chi_B R_{BB})/2R_{AB}$ (8)

Huheey [17] pointed out that the assumption of electronegativity equalization ignores energies arising from electrostatic ("ion-ion") interaction [18] and changes in overlap[19] For bonds which have a high degree of ionic character this is serious, but for predominantly covalent bonds the errors incurred are small. The errors resulting from neglect of changes in electrostatic and overlap terms have opposing effects and tend to cancel each other; both approach zero as δ approaches zero.

2.2. Percentage ionic character:

The most obvious application of electronegativities is the prediction of the polarity of a chemical bond, for which the concept was originally introduced by Pauling. In general, the greater is the difference in electronegativity between two atoms, the more polar is the bond that will be formed between them, with the atom having the higher electronegativity being at the negative end of the dipole. Pauling [1] first assumed that the ionic character may be obtained from the dipole moment of a compound and he compared the dipole moment with the moment produced by the two point charge at a distance apart equal to the internuclear separation of the molecule. And he also made a percentage of the dipole.

The percent of ionic charge was given by-

Percent of ionic charge=$100[1-\exp\{-(1/4)(\chi_A-\chi_B)^2\}]$ (9)

Pauling proceeded to derived the equation for obtaining the dipole charge, plotted these percentages against their electronegativity differences. Thereafter several empirical ansatzs were suggested by various workers to evaluate the dipole charge invoking electronegativities of the bonded atoms from verious scales. The working formulae to evaluate the dipole charge are reviewed below:

(1) Pauling's formula :

Pauling [1] proposed an ansatz to calculate the ionic character of the bond(i.e. static charge) was as

$$q = 1 - \exp(-(\chi_B - \chi_A)^2/4) \qquad (10)$$

where χ_B and χ_A are the atomic electronegativities of atoms B and A respectively. Pauling's method for the calculation of percent of ionic charge was very simple. He estimated the 17 per cent of the moment produced by two point charges at a distance apart equal to the internuclear separation in the hydrogen chloride molecule from the observed dipole moment 1.03 D. Similarly, he estimated per cent of the moments for hydrogen bromide, hydrogen iodide and hydrogen fluoride molecule as 11 per cent, 5 per cent and 60 per cent respectively. This anomalous ionic charge of HF molecule was corrected by Hannay and Smyth, who, [20], measured the dipole moment of hydrogen fluoride molecule and corrected Pauling's value for this molecule as 43 per cent by using a modified percent of ionic charge equation as follows-

$$\text{Per cent ionic character} = 16(\chi_A - \chi_B) + 3.5(\chi_A - \chi_B)^2 \qquad (11)$$

(2) Nethercot's formulae

Nethercot [21] concluded that the dipole charge,q is not only depends on the electronegativity difference of two atoms but also depends on the equalized electronegativity value. He proposed two formulae to calculate the dipole moment charges as under-

$$q = 1- \exp(-3(\chi_B - \chi_A)^2/2\,\chi_{AM}^{\;2}) \qquad (12)$$

$$q = 1- \exp(-(\chi_B - \chi_A)^{3/2} / \chi_{GM}^{\;3/2}) \qquad (13)$$

where χ_{AM} and χ_{GM} are the arithmetic mean(AM) and the geometric mean(GM) of the two atomic electronegativities.

(3) Barbe's formula

Barbe [22] proposed another simple equation to calculate the dipole moment charges of hetero nuclear diatomic molecules. The ansatz which has been proposed by Barbe to calculate the dipole moment charges is as

$$q = (\chi_B - \chi_A) / \chi_B \qquad (14)$$

where $\chi_B > \chi_A$.

(4) Kim's formula:

Using the SBC model and Ray et al electronegativity equalization procedure, Kim [23] calculated the amount of charge transferred, q, for heteronuclear diatomic species as

$$q =\{ r_1r_2/(r_1 + r_2) \}\{(r_BZ_B/r_2r_B) -(r_AZ_A/r_1r_A)\} \qquad (15)$$

Applying the zero-order approximation,i.e. $r_1 \approx r_A$ and $r2 \approx r_B$,Kim obtained

$$q=\{ r_1r_2/CR_{AB}\}(\chi_B - \chi_A) \qquad (16)$$

2.3 Hetaropolar bond length:

Pauling[1] first evaluated the bond length from the electronegativity. The Pauling electronegativity was derived from heats of formation or essentially, bond energies, the electronegativity difference between two atoms reflects the strength of two bonds and moreover there exist a quantitative correlation between electronegativity and bond polarity. On the basis of simple bond charge model(SBC)[14], Ray et al[15] derived the heteropolar bond length, R_{AB}, in terms of the electronegativites, χ_A and χ_B , and covalent radii , $r_A = 1/2R_{AA}$ and $1/2R_{BB}$, of atoms A and B as follows-

$$R_{AB} = (r_A+r_B)-\{(r_A r_B(\chi^{1/2}{}_A - \chi^{1/2}{}_B)^2\}/ (\chi_A\, r_A + \chi_B\, r_B) \qquad (17)$$

2.4. Electronegativity and molecular orbital theory

Coulson[24] introduced the L.C.A.O- Molecular orbital theory. This theory introduced two important quantities

1. the Coulomb integral, $\alpha_r=\int\Phi_r\, \hat{H}\, \Phi_r\, d\tau$, (18)

   which measures the energy of an electron when confined to the atom r within the molecule.

and,

2. the resonance integral, $\beta_{rs}=\int\Phi_r\, \hat{H}\, \Phi_s\, d\tau$. (19)

Coulson et al [25] point out that the Coulomb term can be related (although it is not necessarily) to the ionization potential (I) of atom r. But after a detail examination on this topic, Mulliken[26] concluded that in a bond XY the difference, $\alpha_X - \alpha_Y$, should be proportional to the difference in electronegativity of the atoms X and Y in the proper valence states.

Laforgue [27] and Chalvet and Daudel [28] oversimplified the fact and suggested that, in Pauling scale, $\alpha_X - \alpha_Y = \chi_A - \chi_B$.

2.5.Dipole moment:

Dipole moments μ_d are caused by two opposite charges of magnitude q in Coulombs separated by distance r in meters.

$$\mu_d = q \times r \qquad (20)$$

In the molecular world we use the following form for defining the molecular dipole moment

$$\mu_d = q \times R_{AB}. \qquad (21)$$

where R_{AB}, is the internuclear distance.

Dipole moments are most often expressed in units of Debye where 1 Debye = 3.336×10^{-30} coulomb meters.

As electronegativity is an abstractly defined property and also it is not an observable, hence it cannot be directly measured. However, relative electronegativities can be observed indirectly by measuring molecular dipole moments: in general, the greater the dipole moment, the greater the separation of charges must be, and therefore, the less equal the sharing of the bonding electrons.

It has long been recognized that the dipole moment, μ_d, of the molecule AB can be related to the difference of the two atomic electronegativities ($\chi_B \sim \chi_A$). Indeed in 1932, Pauling[1] proposed the empirical relationship q = 1 - exp [-(χ_B - $\chi_A)^2$] for estimating the ionic character in the molecule. This form was chosen to agree with the then available experimental values of the "dipole moment" charge. In the next year, Malone [29] discovered that dipole moment in Debye (μ_d) of a hetaronuclear bond A-B and the electronegativity difference, χ_A-χ_B, are proportional. i.e.,

$$\chi_A \sim \chi_B \propto \mu_d \qquad (22)$$

In 1935, Mulliken [30] correlated relative electronegativities, effective charges on atoms in partially polar molecule, dipole moments in terms of LCAO-MO coefficients.

He proposed that the LCAO coefficient are affected by polarities of the bond and suggested that the coefficient, 'a' and 'b' are important for the prediction of the electronegativity.

The bonding molecular orbital form for the Linear Combination of Atomic Orbital(LCAO) can be approximately written as–

$$\Phi_{AB} = a\Phi_A + b\Phi_B \qquad (23)$$

If a=b, then the two atoms are same electronegativity, but if a>b then we can say that electronegativity of A >B. Thus, Mulliken[30] pointed out that the difference of electronegativity can be correlated with the difference of the LCAO coefficients, i.e., a-b or more precisely a^2-b^2.

For a diatomic molecule where the bonding electron pair only contributes to the electrical moment the electronic distribution occurs as:

$-2ea^2$ is the charge centered on the atom A,

$-2eb^2$ is the charge centered on atom B.

-4eabS is the charge centered the centers A and B and the distance of the electric center from the midpoint of the bond A-B is considered as z. Then the electrical net moment can be obtained by the contribution of the charges centered in centers A, B, in between A–B and a charge +e on each atom, i.e., $\mu_{Net} = er(a^2-b^2)-4ezabS$, where r is the distance between the atom A and B.

Mulliken[30] defined the term z as the measure of inequality of the polarity of the two atoms A and B. If the molecule AB is homopolar then z=0 its magnitude increases with the inequality of the size of A and B. its sign is such that its positive pole is directed towards the larger atom. Mulliken empirically correlated the charge distribution of the molecule in terms of the electronegativity and proceeds to evaluated the primary dipole moment of the molecule in terms of the LCAO coefficient as-

$(\mu_d)_{AB}= Q_B\, re-(4ezabS-\mu_s)$ (24)

Where Q is the net charge and $Q_A=-Q_B=-e(a^2-b^2)$The term 4ezabS is the Homopolar dipole and Q_B re is the main dipole term.

In case of heteropolar system, this term is cancelled by μ_s

thus for the heteropolar system, we can write

$(\mu_d)_{AB}= Q_B\, r_e$ (25)

Thereafter, in 1954, Dailey and Townes [31] in terms of LCAO coefficients(a and b) proposed that the dipole moment for a heteropolar bond can be expressed as sum of the primary moment(μ_p), overlap moment(μ_o) of the orbitals of atoms , hybridization moment(μ_{hy}) of the valence shell of the atoms and polarization moment(μ_p) of the non bonding electron.

$\mu_d= \mu_p+ \mu_o+ \mu_{hy} + \mu_p$ (26)

They defined the ionic character of a heteropolar bond as the difference between polarizabilities for the electron to be found on atom A or B. They detonated it in terms of the LCAO coefficient as $(b^2{\sim}a^2)$

$\mu_d= eR(b^2{\sim}a^2)+ \mu_o+ \mu_{hy} + \mu_p$ (27)

Dailey and Townes[31] also pointed out that it is not possible to calculate the contribution of the polarization of the non bonding electron to the total dipole moment and considered the earlier assumption (Pauling) that the first term of the equation is of major importance.

i.e., $\mu_d = eR(b^2 \sim a^2)$ (28)

2.6.Atomic Polar Tensor:

Kim [23] extended the SBC model to evaluate the Atomic polar tensor. He showed that electronegativity and the electronegativity equalization can be used as an important tool for the determination of Atomic Polar Tensor in case of diatomic molecule.

Kim[23] proceed to evaluate the dipole charge using the equation (15) and applying the zero-order approximation, i.e. $r_1 \approx r_A$ and $r_2 \approx r_B$

$q=\{ r_1 r_2/CR_{AB}\}(\chi_B - \chi_A)$ (29)

The centroid of positive charger, r, relative to the point defining the centroid of negative charge given by Kim as

$r = \{r_2 Z_B - r_1 Z_A - (r_1 + r_2)\, q\} /(Z_A + Z_B)$ (30)

The dipole moment, μ, was defined by Kim[23] as to direct from the centroid of negative charge to that of positive charge, i.e.,

$\mu = (Z_A + Z_B)r = - (r_1 + r_2)\, q + (r_1Z_B - r_2Z_A)$

$= - R_{AB}\, q + (r_2Z_B - r_1Z_A)$

$= - 1/C\, [(r_Ar_B\, \chi_A\, \chi_B) / (r_A\, \chi_A + r_B\, \chi_B)^2][R^2_{AB}\, (\chi_B - \chi_A)]$ (31)

For a diatomic molecule of AB type where A atom is located at the origin and B atom in positive Cartesian direction and $\chi_B < \chi_A$. Then, the atomic polar tensors(P_x's) for atoms. A and B was given by Kim[23] as

$P_x^B = - P_x^A = (\partial\mu/\partial R)_e$ (32)

Where $(\partial\mu/\partial R)_e$ is the dipole moment derivative at geometric equilibrium.

The simple bond charge (SBC) model [14] for the dipole moment under this condition gives the atomic polar tensor of B atom as follows,

$P_x^B = (\partial\mu/\partial R)_e = - (\chi_B - \chi_A).\, 2R_{AB}r_Ar_BX_A\, X_A\, /C(r_A\chi_A + r_B\, \chi_A)^2$ (33)

Thus one can predict the atomic polar tensors of a molecule AB solely from the electronegativities and covalent radii of the atoms A and B by the Kim formula.

2.7. Bond stretching frequency and Force constant:

Several correlations have been shown between infrared stretching frequencies of certain bonds and the electronegativities of the atoms involved however, this is not surprising as such stretching frequencies depend in part on bond strength, which enters into the calculation of Pauling [1]electronegativities.

The most commonly encountered form of Hooke's law is probably the spring equation, which relates the force exerted by a spring to the distance it is stretched by a force constant, k, measured in force per length.

$$F=-kx \tag{34}$$

Force constant has the general expression-

$$k=-W_1/R^3 \tag{35}$$

Ray et al[15] derived a formula for the computing force constant value form the SBC model as follows-

The vibrational energy function was defined as-

$$W=W_0+(W_1/R)+(W_2/R^2) \tag{36}$$

$$W_1/R=-(Z_A+Z_B)[\{(Z_A+\delta)/r_1\}+\{(Z_B-\delta)/r_2\}-\{(Z_A+\delta)(Z_A+\delta)/R(Z_A+Z_B)\} \tag{37}$$

and,

$$W_2/R^2=h^2(Z_A+Z_B)/8mR^2\upsilon^2_{AB} \tag{38}$$

where W is the Born-Oppenheimer potential, R is the internuclear separation,

The term, W_1/R and W_2/R^2 are assigned to describe [16] the electrostatic energy of the system and the kinetic energy of the bond charge moving freely in a one-dimension box of length υR along the bond respectively.

Using the equalized molecular electronegativity expression, $\chi_{AB}= (\chi_A R_{AA} + \chi_B R_{BB})/2R_{AB}$, they[15] obtained

$$W_1/R=-\{(Z_A+Z_B)^2/R\}[2-\{r_1r_2/(r_1+r_2)^2\}] \quad (39)$$

Ray et al[15] found that the quantity in bracket is close to 7/4 for most of the reasonable values of r_1 and r_2, thus they set it equal to 7/4 to obtain a formula having maximum simplicity. Thereafter using the force constant formula $k=-W_1/R^3$ they obtained

$$k=(7/4)\ \chi_{AB}^2/(R_{AB}C^2) \quad (40)$$

where C is the constant depending on the bond type.

Badger[32] correlated equilibrium bond distance (R) and the bond stretching force constant(k)as

$$K=(c/R)^{1/2}+d, \quad (41)$$

where c and d are the constants.

Reimick[33] pointed out that the bond stretching force constant(k) is a function of product of the electronegativity of the atoms present on the molecule. Thus in a diatomic molecule

$$k_{AB}=f(\chi_A\chi_B) \qquad (42)$$

But the bond stretching force constant (k) is not only a function of product of the electronegativity of the atoms present on the molecule but also it depends on several factors such as (i) bond order(N), and (ii)bond distance(R).

Gordy [34] considered all the factors and modified the Reimick's equation as

$$k_{AB}=f(N,\chi_A\chi_B,R) \qquad (43)$$

Thus he proposed the general equation

$$k_{AB}=a\,N(\chi_A\chi_B/R^2)^{3/4}+b, \qquad (44)$$

where a and b are constants for certain broad classes of molecules. To determine the constants, a and b, Gordy compared calculated $N(\chi_A\chi_B/R^2)^{3/4}$ with the experimental bond stretching values of some molecules and evaluated the constants as, a=1.67 and b=0.30.

It should be notated here that k_{AB} is measured in dynes/cm x 10^{-5}, R is in Å. The value of the constants 'a' and 'b' varies from a class of molecule to another class.

A detail study was made by Gordy and proposed different constant values for different classes and bond order in his paper.

Ray et al[15bm,] on the basis of Pasternak's[16] electronegativity and SBC model[14] showed that the force constant of A-B bond can be directly evaluated from the equalized electronegativity as

$$k_{AB}= (7/4)\,\chi^2_{AB}/R_{AB}C^2 \qquad (45)$$

where, C is a constant which depends on bond type.

2.8. Standard enthalpies of formation and bond dissociation energy:

Bratsch[35] pointed out that the Pauling scale of electronegativity can be used to predict standard enthalpies of formation of binary compounds-

$$\Delta H^0_f =-96.5n\,[\chi_A-\chi_B]^2 \qquad (46)$$

Where, n is the number of equivalents in the compound formula, χ_A, χ_B are the Pauling electronegativity in $(eV)^{1/2}$ unit. ΔH^0_f is the Kilo joule per mole.

Knowledge of bond dissociation energies of chemical bonds in molecules is essential for understanding chemical processes [36]. At the outset it should be emphasized that there is a distinction [37] between bond dissociation energy and average bond energy. The bond dissociation energy of a bond in a molecule A-B is the energy needed to separate the radicals A and B to infinity, each species being in its ground state. The average bond energy of a bond A-B is defined as l/nth the energy needed to separate each of the atoms in a symmetrical molecule AB, to infinity, all species being in their ground states, that is, l/nth of the heat of atomization of the molecule. In general, the bond dissociation energy plays a more important role in chemistry than the average bond energy. Both Szwarc[37] and Steacie [38] have demonstrated the importance of bond dissociation energies in the interpretation of chemical kinetic data. They[37-38] have shown how the analysis of complex reactions into their elementary reactions can be aided greatly by the knowledge of the bond dissociation energies. Because thermal, photochemical, radiation and discharge reactions are usually complex, tables of bond dissociation energies should be very useful in interpreting kinetic data in these fields. Bond dissociation energies also can be employed to calculate thermochemical properties. Heats of formation of radicals can be obtained from bond dissociation energies, which, in turn, can be employed for calculating heats of reactions involving free radicals. Furthermore, bond dissociation energies can be used to calculate heats of reactions in which free radicals are not involved.

Finally, knowledge of bond dissociation energies is essential in interpreting most results obtained by electron bombardment of molecules. In a mass spectrometer one measures the appearance potential of an ion of known mass. In order to deduce what process occurred in the mass spectrometer, one calculates the appearance potentials of various possible processes and by comparing those to the observed value the most likely process can be chosen. To calculate the appearance potentials knowledge of the bond dissociation energies is necessary.

A discussion of the use of bond dissociation energies for interpreting electron impact data has been given by Fineman and Martins[39] and Field and Franklin[40].

The Bond dissociation energies or the bond energy (BE) for a bond A-B is defined as the standard-state enthalpy change for a reaction of the type $AB \rightarrow A+B$ at a specified temperature (T).

$$(BE)_T = \Delta H_f^0(A) + \Delta H_f^0(A) - \Delta H_f^0(AB) \qquad (47)$$

where ΔH_f^0 is the standard state heat of formation.

The earliest method which correlate the bond energy with electronegativity is the landmark Pauling's equation[1] which is based on single bond energy. The scale correlates the extra ionic resonance energy to the electronegativities of atoms. The energy difference Δ, was defined by-

$$\Delta = D(A\text{-}B) - (1/2)[D(A\text{-}A) + D(B\text{-}B)]. \qquad (48)$$

Pauling was able to assign electronegativity of many elements which roughly satisfy the equation

$$\Delta = (\chi_A - \chi_B)^2 \qquad (49)$$

Pauling took, $\Delta = -\Delta H_f^0$(heat of formation)

Within the framework of SBC model [14], Pasternak [16] pointed out that in the limit of infinite internuclear separation, a diatomic molecule becomes two ions at a infinite separation and a bond charge at an infinite distance from the ions and free to extend over an infinite volume. The energy of this system from the equation $W=W_0+(W_1/R)+(W_2/R^2)$ is $W(\infty)=W(0)$, but the W(0) value is arbitrary. The real end product of the dissociation of a homonuclear diatomic molecule is of course two neutral atoms at infinite separation. Therefore, having the $R=\infty$ configuration of the ions and the bond charge, Pasternak then divided the bond charge and recombined it with the ions. This involves the gain in energy by an amount W_i , equal to the twice of the energy required to removal of Z electrons from a single atom.

Thus, the energy of the system of two infinitely separated systems of two infinitely separated neutral atoms is zero.

Thus, $W(0)=W_i$ (50)

The dissociation energy D is-

$D=W(\infty)-W_i-W(R)$ (51)

Or, $D=-W_i-(W_1/R)-(W_2/R^2)$ (52)

Pasternak[16] found that the term W_i is continuous function of Z as–

$W_i=2(aZ+bZ^2)$ (53)

where the coefficients a and b are determined from the experimental ionization energies.

Pasternak defined the electronegativity in the SBC framework as Z_A/r_A where Z_A is the effective nuclear charge and r_A is the covalent radius of atom A in a molecule.

Although many bond dissociation energies of chemical bonds between various kinds of atoms in many molecules have been determined experimentally, there is at present no reliable theoretical or empirical method for estimating bond dissociation energies.

2.9. Stability ratio:

The ratio of the average electronic density of an atom to that of a hypothetical, isoelectronic inert atom, is termed[41] as the "stability ratio" (SR)

Sanderson[41] defined the term 'stability ratio' (SR) as–

$SR = D/D_i$ (54)

with $D = 3Z/4\Pi r^3$ (55)

D_i was defined as the electron density of an isoelectric inert atom, determined by interpolation between real values, which was needed to correct the average electron density (D) for variations in Z that were unrelated to chemical reactivity. Previously Sanderson [42] pointed out that the relative average electronic densities of the atoms of the elements can be used as a tool for the measurement of the electronegativity. The stability ratio thus can be estimated from the knowledge of electronegativity. The relation of stability ratios to Pauling electronegativities [1] has previously been presented graphically [41] , but the numerical dissimilarity has made difficult a precise quantitative comparison. Accordingly, a simple conversion from SR values to the arbitrary units of Pauling's electronegativity is presented here.

An empirical mathematical relationship is:

$$\chi^{1/2} = 0.21SR + 0.77 \quad (56)$$

where χ is the electronegativity in Pauling scale.

He also found that the stability ration of an atom is a measure of its electronegativity. Thus,

$$SR = \chi_{Sanderson} \quad (57)$$

Sanderson found that halogen atoms bear a linear relationship between electronegativity and experimental electron affinities. He also established a linear relationship between acid/base strength and SR for that particular atom.

Sanderson then defined the partial charge of an atom as a function of the difference between the elemental electronegativity and that of the new electronegativity, the equilibrium electronegativity. His original equation dealt with the difference between the stability ratio (SR) for the entire molecule and the SR for the atom for which the partial charge is being calculated.

2.10. Lewis acid strength:

Smith[43] has complied a numerical scale of acid base character for binary oxides by analogy with the Pauling electronegativity scale. For such oxides he has evaluated an acid base parameter ‘a’.

Bratsch[44] point out that this parameter can be used to predict the standard enthalpy of combination of binary oxide to form oxo-salt as-

$$\Delta H^0_{comb} = -[\ a_A - a_B]^2 \tag{58}$$

where a_A and b_B are the Smith acid base parameter.

Bratsch[44) suggested that the equalized χ_{eq} in an oxide may be estimated by $N/\Sigma[v/\chi]$ where N is the number of atoms in the oxide formula, v is the number of each element in the oxide formula, χ is the initial pre bound electronegativity of each element on the pauling scale.

Bratash[45] also gave a linear relationship between Smith oxide acid base scale and Pauling electronegativity as

$$a = m\ \chi_{eq} + b \tag{59}$$

$$a = 9.8\ \chi_{eq} + b \tag{60}$$

The potential change δ_0 on combined oxygen was given by Bratash as follows–

$$\delta_0 = (\chi_{eq}-\chi_0)/\chi_0 \qquad (61)$$

Thereafter he correlated the Smith acid base parameter with the partial charge as

$$a = 33.7\ \delta_0+9.2 \qquad (62)$$

Brown and Skowron[46] proposed a scale to measure the Lewis acid strength(S_a) as

$$S_a = V/N_t \qquad (63)$$

where V is the oxidation state of the cation and N_t is the average of the coordination number to the oxygen observe

They suggested that the average Lewis acid strength is the Pauling bond strength averaged over all the compounds in which the cation appears.

They relates S_a and electronegativity χ as

$$Sa = 1.18\chi^2 \qquad (64)$$

where χ is in Rydbergs unit and S_a is in valence unit(vu)

2.11. Electronegativity and the work function:

The work functions of metals, the electrode potentials of elements, are also related to electronegativity. The work function of a metal is the minimum work required to remove an electron at 0K. Gordy and Thomas [47] made an attempt to relate the two parameters and presented an empirical relation between work function Φ and electronegativity χ as

$$\chi(\text{Pauling})=0.44\ \Phi - 0.15 \quad (65)$$

Subsequently, Conway and Bockris[48], Miedema et al[49] and Trasatti [50]modified the electronegativity-work function relation of Gordy and Thomas. All used the same straight line relation proposed by Gordy and Thomas. All the above workers used Pauling scale to correlate electronegativity and work function but Michaelson [51] made an exploratory test of the relation between Mulliken electronegativity[52] and the work function. He proposed a general equation

$$\Phi= \chi_{\text{Mulliken}}-P \quad (66)$$

Michaelson called the term P as periodicity parameter; a quantity which is a measure of the difference between atomic and solid state periodicity.

2.12. Calculation of other periodic parameters:

Although the periodic table does not directly follow from the quantum mechanics, it has the powerful chemical organizing power. The development of modern chemistry began with the finding of periodic changes in properties of elements leading to the formulation of the periodic table as well as periodicity law of elements.

The concept of the electronegativity can be employed for the prediction of the stability of chemical bonds, in the calculation of the surface atoms properties, reaction enthalpies etc side by side based on the periodic law, we can calculate one periodic parameter from another. Another parameter which defines the space occupied by an atom or an ion is its radius, whose value can be estimated by various theoretical methods. Although it does not have any precisely defined physical sense, it allows us to formulate a useful working hypothesis, often employed in solutions of various problems in chemistry, physics, biology and others. The periodic table tells us that the electronegativity and radius are inversely related to each other[53]. Ghosh [53] offered a scale of electronegativity based on the absolute radius of the atom where it is stated that the electronegativity and atomic radius(r) are inversely proportional to each other. He suggested a linear relationship between χ and r.

Another very old and one of the most useful periodic parameter is the hardness. The notion of hardness was first introduced by Mulliken [54] when he pointed out that the 'Hard' and 'Soft' behavior of various atoms, molecules and ions can be conceived during acid-base chemical interaction. Soon after Mulliken's classification, the terms hardness and it's inverse, the softness were in the glossary of conceptual chemistry and implicitly signified the deformability of atoms, molecules and ions under small perturbation. The hardness refers to the resistance of the electron cloud of the atomic and molecular systems under small perturbation of electrical field. An atom or molecule having least tendency of deformation are hard and having small tendency of deformation are soft. In other words, list polarizable means most hard and in such systems the electron clouds are tightly bound to the atoms or molecules. On the contrary most polarizable means least hard and in such systems the electron cloud is loosely bound to the atoms or molecules. Electronegativity though defined in many different ways, the most logical and rational definition of it is the electron holding power of the atoms or molecules [7]. The more electronegative species hold electrons more tightly and the less electronegative species hold less tightly. Thus, if we invoke the qualitative definition of hardness stated above and compare with the qualitative definition of electronegativity, the commonality of their conceptual structures and philosophical basis are self-evident.

Ghosh and Islam [7] recently pointed out the commonality in the fundamental nature of hardness and electronegativity– the holding power of the electron cloud by the chemical species. Thus the qualitative views of the origins of hardness and electronegativity nicely converge to the one and single basic principle that they originate from the same source –the electron attracting power of the screened nuclear charge [7]. As it is a fact that the origin and the operational significance of the electronegativity and hardness are the same, we may conjecture that the two periodic parameters are directly related to each other.

Putz[55], after rigorous research on electronegativity and hardness, opined out that the hardness and electronegativity are proportional to each other, i.e., $\chi \propto \eta$.

Ayers[56] proceed to evaluate the electronegativity and hardness of neutral atoms on the basis of the energy expression of March and White[57] and pointed out that the two fundamental atomic parameters, hardness and electronegativity, have the similar expression.

Xue et al [58] suggested that the hardness of atoms can be defined as the electron holding energy of atoms per unit volume i.e., $\eta_a = \chi_a/r^3$.

One of the values characterizing atoms and free ions is their ionization energy, a value that determines the energy with which an electron interacts with the atomic core containing a nucleus and the remaining electrons. Another similar value is the electron binding energy in the outermost filled electronic shell in atoms of elements in their thermodynamically stable forms. This energy characterizes the atom core, the 'ion core' with valence electrons removed, in a state in which the atoms form metallic bonding in metals or atomic bonding. Progresses in spectroscopic methods the above periodic parameters have an important feature of being experimental values. It is not derived from any models and due to advances in spectroscopic methods is determined accurately.

Mulliken[52] electronegativity scale correlated the three periodic parameters together as

$$\chi = (I+A)/2 \qquad (67)$$

Thus the any one among the three periodic parameters can be calculated by the knowledge of other two periodic parameters.

The valence electron density is a parameter which can define and measure the electronegativity of an atom. Nagle [59] found that the cube root of this ratio of the number of valence electrons divided by the polarizability, $(n/\alpha)^{1/3}$, can be used as a measure of electronegativity for all s- and p-block elements (except the noble gases). Ghosh and Gupta [60] also suggested that both polarizability and electronegativity are periodic properties and they are connected by an inverse relationship, $\chi \propto (1/\alpha)^{1/3}$.

2.13. Electronegativity and the HSAB Principle:

One of the very purposes of the modern conceptual chemistry stands the capacity of modeling and controlling the chemical reaction via theoretical methods. There is also recognized that only with admission of the electronegativity and hardness concepts in the chemical reactivity principle, like Hard and Soft Acid and Base principle has the benefit to describe the several fundamental phenomena on a sound mathematical way. The Hard and Soft Acid and Base is a rule of thumb proposed by Pearson [61] in 1963 in order to generalize the acid base interactions. This principle was empirical till the publication of the landmark paper by Parr and Pearson[62]. In this study, based on the assumption that the energy is quadratic and utilizing the concept of absolute hardness, Parr and Pearson[62] presented a theoretical deduction of the HSAB principle. They assumed the formation of A: B from A and B: may be regarded as comprising two components: (i) shift of some charge, ΔN from B to A and (ii) formation of the actual chemical bond.

Focus primarily on the first effect, they wrote the energy expression for A and B in the molecule as follows-

$$E_A = E_A^0 + \mu_A^0(N_A - N_A^0) + \eta_A(N_A - N_A^0)^2 \qquad (68)$$

and

$$E_B = E_B^0 + \mu_B^0(N_B - N_B^0) + \eta_B(N_B - N_B^0)^2 \qquad (69)$$

The electron numbers, N_A and N_B are

$$N_A = N_A^o + \Delta N \quad (70)$$

and

$$N_B = N_B^o - \Delta N \quad (71)$$

After the formation of the molecule AB, the chemical potentials of A and B are equal in the molecule (electronegativity/chemical potential equalization principle). Thus

$$\mu_A = \mu_A^o + 2\eta_A \Delta N = \mu_B = \mu_B^o - 2\eta_B \Delta N \quad (72)$$

The shift of charge / electron transfer,

$$\Delta N = (\mu_B^o - \mu_A^o)/2\,(\eta_A + \eta_B) \quad (73)$$

as, $\chi = -\mu$

Thus,

$$\Delta N = (\chi^0_A - \chi^0_B)/2(\eta_A + \eta_B) \quad (74)$$

The corresponding energy change was calculated as follows-

$$\Delta E = (E_A - E^0_A) + (E_B - E^0_B) \quad (75)$$

$$= -(1/2)(\mu_B^o - \mu_A^o)\,\Delta N$$

or,

$$\Delta E = -(\chi^0_A - \chi^0_B)^2/4(\eta_A + \eta_B) \qquad (76)$$

As the acid must be more electronegative than base, $(\chi^0_A - \chi^0_B)$ is always positive, an energy lowering results from electron transfer process. The difference in absolute electronegativity drives the electron transfer, and the sum of the hardness parameter acts as a drag or resistance. In other words, the differences in electronegativity drive the electron transfer and the sum of the absolute hardness parameters inhibits electron transfer.

If both acid and base are soft, $(\eta_A + \eta_B)$ is a small number, and for a reasonable difference in electronegativities, ΔE is substantial and stabilizing. This explains the HSAB principle, meanwhile, it seems safe to say that it explains a part: soft prefers soft. But if both acid and base are hard, there is little electron transfer and energy stabilization from electron transfer, for a given difference in electronegativities. Parr and Pearson[62] commented- "*This result seems paradoxical*" and there is the need of the second effect- the formation of the chemical bond.

Parr and Pearson[62] also commented that the consideration "*soft-soft interactions are largely covalent, and that hard-hard interactions are largely ionic*" is not always novel.

Two years afterward, Pearson [63] explained the adduct formation between a neutral acid and a neutral base. Providing η_A and η_B are both small, the stabilization of A:B adduct can be explained by double bonding. The concept of double bonding resembles with the π-bonding theory of Chatt[64]. Chat and coworkers[64] used it for explaining various metal ion-ligand preferences. In case of the adduct formation between a hard acid and a hard base normally "little two-way electron transfer" occur. It should be notated that Pearson also showed that there will be little one-way transfer from B to A, if η_A and η_B are large. For cationic acids the probability of double bonding is greatly reduced. The main source of bonding will come from ionic bonding or ion-dipole bonding. Neutral molecules are the most likely to have two-way electron transfer. The unbiased values of $(\chi^0_A - \chi^0_B)$ for the neutral molecule determine the direction of net electron transfer. The total amount of electron transfer is governed by $(\eta_A + \eta_B)$ and a small value of the summation is favorable for maximum covalent bonding.

The concept of electronegativity provides a measure of the intrinsic strength of an acid or base [56, 65-66]. A strong Lewis acid is a good electron acceptor and has high electronegativity/low chemical potential. A weak Lewis acid has a lower electronegativity than a strong Lewis acid, but a higher electronegativity than a Lewis base. A strong Lewis base readily donates electrons and has a lower electronegativity than a weak Lewis base. These relations are summarized by Ayers [56]as follows-

χ(strong acid)> χ(weak acid)> χ(weak base)> χ(strong base)>0 (77)

The perfect electron donor has $\chi = 0$. One can reify the electron-accepting abilities of real molecules by imagining how they would react with a perfect electron donor. Thus the electronegativity concept plays a dominating role in the principle of Hard and Soft Acid and Base.

One of the most important questions connected with the problem of reactivity of molecules in different environmental conditions is the prediction and interpretation of the preferred direction of a reaction and the product formation. Sekhon[67] examined metathesis reaction of the type AB+CD=AD+BC in terms of the equalized electronegativity values of various species involved in the double exchange reaction. The conclusion was made by him from the study that an exchange reaction proceeds from left to right if the total sum of the equalized electronegativity value of the products is greater than that of reactants.

2.14. The concept of Group electronegativity:

One important application of the electronegativity concept is in the estimation of the electron-withdrawing ability of chemical groups. For several years' scientists, especially who specialize in the area of synthesis; have expressed a desire for a group electronegativity scale. This application requires the ability to account for charges on group. The idea of group electronegativity is important because the electronegativity concept evolved largely from the desire of organic chemists to understand reaction mechanism in terms of the inductive effects of various functional groups.

Meek et al[68] extended the original Pauling concept of electronegativity for defining group electronegativity as the Power of a group in a molecule to attract electron to itself. They pointed out that the groups have a better ability to donate or accept charges than atom and therefore be considered as reserver of enhanced charged capacity. Hence a group of atoms as a unit is potentially able to donate or withdraw considerable amounts of charge with a very little effect on itself. The ability to dissipate charge over several atoms increases as the number of atoms which constitute the group increases. Upon bond formation between two atoms, charge is transferred from one atom to the other. In the case of a chemical group charge is transferred to the central atom because of its bond. In CH_3, for example, the charge is transferred to the C atom because of the CH bonds. The electronegativity of the group will then be the orbital electronegativity of the central atom suitably modified to account for its charge. [69]

2.15. Some other applications of electronegativity:

This section deals with some other methods of assessing electronegativity that have been proposed but which, in the view of the scientists, are rather less acceptable than those described so far. The electronegativity concept when correlated with other atomic parameters, it become useful to explain various complicated phenomena such as crystal structure, electron structure, non linear optical polarizability, ultraviolet reflection coefficient and the valence bond photo emission[70].

The concept of electronegativity can be used in the correlation of the chemical shifts in NMR spectroscopy or isomer shifts in Mössbauer spectroscopy. More convincing are the correlations between electronegativity and chemical shifts in NMR spectroscopy. [71] or isomer shifts in Mössbauer spectroscopy[72]. Both these measurements depend on the s-electron density at the nucleus, and so are a good indication that the different measures of electronegativity really are describing "the ability of an atom in a molecule to attract electrons to itself". A Systematic correlation between electronegativity and values of Tc for superconductive elements, binary alloys and high Tc-oxides was reported [76-74]. Superconductive elements have electronegativity values that are concentrated within a range near the center of Pauling's scale (1.3-1.9).On the same scale they lie between 2.5 and 2.65 for the high Tc-oxides. Ichikawa found a trend where the Tc values for some high Tc-oxides increase systematically with weighed harmonic mean electronegativity (χ_{WH}) in such a way that they head towards some optimum value which lies between 4.69 and 4.88 eV. Furthermore a wide range of phenomena [5] such as ligand field stabilization, the first filling of p orbitals, the transition-metal contraction and lanthanide contraction can be understood in terms of electronegativity. The refractive index of silica polymorphs is related with average electronegativity. A large average χ i.e., compact electron cloud will give rise to small polarizability and hence small refractive index. The concept of electronegativity is exclusively used in the coordination chemistry. The electronegativity is a measure of the chemical reactivity of an atom, ion, radical, or molecule, which gives the direction of the electron flow an estimate of the initial amount of the charge transferred which is closely related to the energy barriers for reaction and to the strength of the co-ordinate bond formation.

The scale and concept of electronegativity is applied for exploring the structure-property relationship of materials. The construction of the structure activity property relationship based on the electronegativity to predict the properties of materials and further design new materials is a major trend in the development of the electronegativity. Rare earth luminescent materials have been intenshively investigated due to their applications in the field of lasers and medicine [70].

The concept of electronegativity is also very important in the research of rare earth material [70] because both the valence charge and the charge transfer energy are related to the electronegativity.

Devautour et al[75] proposed a new interpretative method for thermally stimulated depolarization current (TSDC) measurements. By applying the concept of electronegativity equalization to TSDC results, Devautour et al[75]showed that it is possible to obtain an experimental evaluation of the chemical potential of the electrons of the exchanged cation and of the host site in zeolites. This step leads to an evaluation of the fundamental parameters, such as the effective hardness and electronegativity of the sites of zeolites. Such an approach gives an evaluation of the heterogeneity of the aluminosilicate surface and is applied to an exchanged hydrogen mordenite containing various amounts of substituted lithium ions or sodium ions.

Schaeffera et al[76] have applied the empirical relationship between electronegativity and effective work function to a diverse set of multi-element electrode materials on hafnium dioxide (HfO_2) gate dielectrics. To accommodate the multi-element nature of metal gate electrodes the group electronegativity of the metal was calculated from the geometric mean of electronegativity with respect to the volume stoichiometry of the constituent elements. Their finding suggested that the group electronegativity concept is also extended to work function engineering via dielectric capping materials. The electronegativity trends provide insight into the relative charge neutrality levels of candidate dielectric capping materials and their subsequent impact on the metal effective work function. Baeten and

Geerlings[77] used the electronegativity equalization principle to study the charge distributions in enzymes. Ramsden [78] studied the influence of electronegativity on the triangular three-centre two-electron bonds. Recently Reddy et al [79] showed the correlation between the optical electronegativity and the refractive index of ternary chalcopyrites, semiconductors, insulators, oxides and alkali halides. Douillard et al [80] obtained the solid surface tension of ideal crystals of talc and chlorite. From this result, it is possible, using thermodynamic models, to calculate the heat of immersion in water of these solids and to compare with experimental data obtained for well-known samples. Their study confirmed that the differences between surfaces of talc and chlorite and confirming that a route of calculation of surface tension using electronegativity equalization is very simple and correct. Kwon et al [81] opined that the electronegativity and chemical hardness are the two helpful concepts for understanding oxide nanochemistry. Makino [82] pointed out that the band gap, heat of formation and structural mapping for sp-bonded binary compounds can be interpreted on the basis of bond orbital model and orbital electronegativity

A relationship between dehydroxylation temperature and electronegativity was suggested by Ray et al [83]. The relationship between the electronegativity and the charge-injection barrier at organic/metal interfaces was suggested by Tang et al [84]. Portier et al [85] studied the exclusive role of electronegativity in the materials design. Now a day, the electronegativity equalization methodology, EEM,[86-89] is frequently used to calculate the charge distribution and reactivity index e.g., local softness and hardness[90-91], condensed Fukui function[92-93], electrophilicity index[94-96] of molecules. Chemical Reactivity Theory (CRT) contains reactivity indices defined as first and second derivatives of ground-state properties with respect to electron number such as the electronegativity and the hardness[55,97].

The study of electronegativity is now an animated field of current research [5]. We may mention some earlier review works [4-5, 98-100] in various field of chemistry which clearly indicate the fact that the applications of the electronegativity is of great theoretical as well as practical interest.

Conclusion:

The electronegativity is very insightful concept in the theory of chemical bonding explaining the formation, stability and the structure of molecules and it is quite natural that it assumes a finer structure as the concept of valence bonding develops and evolves with time. Some very fundamental quantities of both inorganic, organic, and physical chemistry are the concept of bond energies, polarities, and the inductive effects, etc can only be conceived in terms of electronegativity. As the electron density and the effective applied potential function are closely related with the shell structure of atoms, the present study conclude that the electronegativity concept appear to generate the minimal and sufficient set of global parameters that assists the chemical bonding and reactivity, in various chemical and physical conditions. However, the attempts to refine electronegativity theory are not yet sufficiently complete to enable a verdict to be reached on their efficacy; one might reasonably expect that a clearer understanding of electronegativity property will be gained with their further development.

Reference:

[1] L. Pauling, The Nature of the Chemical Bond.3rd edn.;Cornell University:Ithaca.NY .(1960), (b) ibid, J. Am. Chem. Soc., 54, (1932), 3570. (c) L. Pauling, D.M. Yost, Proc. Nat. Acad. Sci., 18, (1932),414.

[2] L Allen J. Am. Chem. Soc. 114, (1992), 1510.

[3] N.Islam, S.Biswas, C.C.Ghosh, Coord Chem. Rev,2009 communicated(MMs no.-CCR-D-09-00112).

[4] H.O Pritchard.; H.A. Skinner, Chem. Rev. 55, (1955), 745.

[5] D.C Ghosh, J. Indian Chem. Soc, 80,(2003), 527.

[6] C.A. Coulson,Proc.R.Soc.London Ser.A., 207 (1951) 63.

[7] D.C. Ghosh, N. Islam, Int. J. Quantum Chem. 109,(2009), in press.

[8] R. T. Sanderson, Science, 114, (1951) 670.

[9] R. G., Parr, R. A. Donnelly, M. Levy, W. E. Palke, J. Chem. Phys, 68, (1978), 3801.

[10] R.G Parr, L.J. Bartolotti J. Am. Chem. Soc 104, (1982), 3801.

[11] R.G. Parr , W.Yang, Density Functional Theory of Atoms and Molecules, Oxford University Press, 1989.

[12] P.Hohenberg , W.Kohn. Phys. Rev. B, 136, (1964), 864.

[13] E.P. Gyftopoulos, G.N. Hatsopoulos, Proc. Natl. Acad. Sci. US, 60, (1965),786

[14] (a) R.G.Parr, R.F. Borkman, J Chem Phys 46, (1967), 3683, (b)R.F. Borkman, R.G. Parr ,J Chem Phys 48, (1968),1116, (c) R.G. Parr, R.F. Borkman, J Chem Phys 49, (1968),1055.

[15] N.K. Ray, L. Samuels, R.G. Parr, J Chem Phys70, (1979), 3680.

[16] A.Pasternak, Chem.Phys., 26, (1977), 101.

[17] J. E. Huheey, J.Phys. Chem.,69, (1965), 3284.

[18] R.P.Iczkowski,J Am. Chem.Soc.,86, (1964)2329.

[19] H.P.Pritchard, J Am. Chem.Soc.,85, (1963), 1876.

[20] N.B.Hannay, C.P.Smyth J. Am. Chem. Soc 68, (1946),171.

[21] (a) A.H. Nethercot, Jr. Chem Phys Lett 59: (1978),346, (b) ibid, Chem Phys 59, (1981),297.

[22] J.Barbe, J Chem Educ 60, (1983)640.

[23] K.Kim, Bull K Chem Soc 8, (1987),432.

[24] C.A.Coulson, Dictionary of Values of Molecular Constants, edited by C. A. Coulson and R. Daudel, Vol. 1. Centre de chimie theorique de France, Paris, 1953.

[25] C. A. Coulson, C. Longuet-Higgnhs, Proc. Roy. Soc. (London) A191, (1947),39.

[26] R.S. Mulliken, Phys. Rev. 74, (1948),736.(85) J. Chim. Phys. 46, (1949),497.

[27] A.Laforguae J. Chim. Phys. 46, (1949), 568.

[28] O. Chalvet, R. Daudel, J. Chim. Phys. 49, (1952),77.

[29] J.G. Malone, J. Chem.Phys, 1, (1933)197.

[30] R.S. Mulliken J Chem Phys, 3, (1935), 573.

[31] B.P. Dailey, C.H. Townes, J. Chem.Phys, 23, (1955),118.

[32] R.M.Badger, J Chem Phys,2, (1934),128.

[33] A.E.Remick, Electronic Interpretation of Organic Chemistry, John Wiley & Sons, Inc., NY,1943.

[34] W. Gordy, J Chem Phys,14,(1946),305.

[35] S.G. Bratsch, J Chem Edu,65,(1988),877.

[36] M.A. Fineman J. Phys. Chem., 62, (1958), 947.

[37] M. Szwarc, Chem. Reus., 47, (1950),75.

[38] E. W. R. Steacie, "Atomic and Free Radical Reactions," Vol. I, Reinhold Publ. Corp., NY,(1954).

[39] M. A. Fineman,G. F. Martins, AEC Report No. NYO 7237, Nov. 1955.

[40] F. H. Field, J. L. Franklin, "Electron Impact Phenomena and the Properties of Gaseous Ions." Academic Press, Inc. New York, N. Y., 1957.

[41] R.T.Sanderson, J. Chem. Educ.29, (1952),539.

[42] R.T.Sanderson, J Chem Phys, 74,(1951),1702.

[43] D.W.Smith, Journal of Chemical Education, 64 (1987), 480.

[44] (a)S.G. Bratasch J Chem Edu, 61, (1984),588 (b) ibid, 62, (1985)101

[45] S.G.BratashJ Chem Edu, 65,(1988),877.

[46] I.D.Brown, A.Skowron, J Am Chem Soc, ,112, (1990), 3403.

[47] W.Gordy, W.J.Orville Thomas, J.Chem.Phys, 24,(1956),439.

[48] B.E.Conway,J.O'M.Bockris, J.Chem.Phys,26, (1957), 532.

[49] A.R.Miedema, F.R.de Boer, P.F.de Chatel, J.Phys. F:Metal Phys.,3, (1973),1558.

[50] S.Trasatti, J.Chem Soc., Faraday Trans. 168, (1972),229.

[51] H.B.Michaelson, IBM, J. Res. Develop. 22,(1978),72.

[52] R. S. Mulliken, J. Chem. Phys., 2, (1934), 782.

[53] D.C.Ghosh, J Theor Comput Chem 4,(2005), 21.

[54] R.S.Mulliken, J. Am. Chem. soc, 74, (1952),811.

[55] M.V.Putz, Absolute and Chemical Electronegativity and Hardness, Nova Science Publishers, Inc.,New York, 2008.

[56] P.W.Ayers, Faraday Discuss, 135, (2007),161.

[57] N.H.March, R.J.White, J.Phys.V, (1972),5.

[58] K.Li, X.Wang, F.Zhang, and D.Xue, Phys Rev Lett, 100, (2008),235504.

[59] J.K. Nagle, J. Am. Chem. Soc., 112, (1990), 4741.

[60] D.C. Ghosh, ; K.Gupta, , J. Theor Comput Chem, 5, (2006), 895.

[61] R.G. Pearson, J Am Chem Soc, 85, (1963), 3533; (b) R.G Pearson, Science, ,151, (1966),172.

[62] R.G Parr. R.G. Pearson, J. Am. Chem. Soc. 105, (1983), 7512.

[63] R.G.Pearson J. Am. Chem. Soc. 107, (1985), 6801.

[64] S.Ahrland, J.Chatt, N. R.Davies, Q. Reu. Chem. Sot. 12, (1958), 265. Chatt, J. J. Inorg. Nurl. Chem. 8, (1958), 515. Ahrland, S. Slrucr Bonding (Berlin) 1, (1966), 207.

[65] N.C.Baird, M.A.Whitehead, Theor. Chim. Acta,2, (1969), 259.

[66] A.R.Orsky,.M.A.Whitehead, Can. J. Chem,65,(1987),1970.

[67] B.S. Sekhon,Proc Indian Natl Sci Acad A, 64,(1998), 581.

[68] Leah D Garner-O' Neale; Alcindor F Bonamy, Terry L Meek : Brian G Patrick, J Mol Struct(Theochem,) 639,(2003),151.

[69] J.Mullay J Am Chem Soc 106,(1984),5842.

[70] L KeYan, X DongFeng,Chinese Sci Bull,54,(2009),328.

[71] H Spieseke; W G Schneider, J. Chem Phys 35: (1961), 722.

[72] C A Clasen, M L Good, Inorg. Chem. 9, (1970), 817.

[73] S Ichikawa, J. Phys. Chem.93 (1989),7302.

[74] D. E. Ramaker ,J Electron Spectroscopy and Related Phenomena, 52, (1990), 341.

[75] S. Devautour, J. C. Giuntini, F. Henn, J. M. Douillard, J. V. Zanchetta, J. Vanderschueren J. Phys. Chem. B, 103, (1999),3275.

[76] J.K. Schaeffera, D.C. Gilmera, C. Capassoa, S. Kalpata, B. Taylora, M.V. Raymonda, D. Triyosoa, R. Hegdea, S.B. Samavedama, B.E. White Jr, Microelectronic Engineering, 84,(2007), 2196.

[77] A. Baeten, P. Geerlings, J Mol Struct (Theochem), 465, (1999), 203.

[78] C A. Ramsden, Tetrahedron,60, (2004), 3293

[79] R.R. Reddy, K. R. Gopal, K. Narasimhulu, L. S S. Reddy, K. R Kumar, C.V. K. Reddy, S. N. Ahmed, Opt. Mat., 31,(2008),209

[80] J. M. Douillard, F. Salles, M.Henry, H. Malandrini, F. Clauss, J Col Int Sci, 305, (2007), 352.

[81] C. -W. Kwon, A. Poquet, S. Mornet, G. Campet, M. -H. Delville, M. Treguer, J. Portier Materials Letters, 51,(2001), 402.

[82] Y. Makino, Intermetallics, 2, (1994), 55.

[83] L.F. Ray , L E Kristy , Matt L. Weier, A.R. McKinnon, P. A. Williams, P. Leverett, Thermochimica Acta, 427, (2005), 167

[84] J.X. Tang, C.S. Lee, S.T. Lee, Y.B. Xu, Chem Phys Lett, 396, 2004,92

[85] J. Portier, G. Campet, J. Etourneau, M.C.R. Shastry, B. Tanguy, J Alloys Compd, 209,(1994), 59

[86] W.J. Mortier; K. Van Genechten, J Gasteiger, J. Am. Chem. Soc. 107, (1985), 829.

[87] W.J.Mortier,S.K. Ghosh, S. Shankar, J. Am. Chem. Soc.108, (1986) 4315.

[88] K.A. Van Genechten, W.J. Mortier, P.Greelings, J. Chem. Phys. 86, (1987), 5063.

[89] R Svobodová Vařeková , Z. Jiroušková , J. Vaněk , Š. Suchomel ,J. KočaInt. J. Mol. Sci. 8, (2007) 572.

[90] M. Berkowitz, R.G. Parr,J Chem Phys, 88,(1988)2554.

[91] A K. Chandra, M T Nguyen Int. J. Mol. Sci. 3, (2002),310

[92] R.G. Parr, W. Yang, J. Am. Chem. Soc. 106 (1984) 4049.

[93] K. Fukui, Science 218 (1982) 747.

[94] R G. Parr, L. v. Szentpály,S Liu,J. Am. Chem. Soc., 121,(1999),1922..

[95] P. K. Chattaraj, U. Sarkar, D. R. Roy, Chem. Rev., 106 ,(2006), 2065.

[96] P.K. Chattaraj ,D. R. Roy, Chem. Rev., 107 (2007),PR46.

[97] M H Cohen,.A. Wasserman, J. Stat. Phys., 25,(2006) 1121.

[98] K.D. Sen, C. K. Jorgensen Electronegativity ,Springer Verlagm NY,1987.

[99] AR Cherkasov, V I Galkin, E M Zueva, R A Cherkasov, Russ. Chem. Rev. 67, 1998, 375.

[100] R G. Pearson, J. Org. Chem., 54 (1989),1423.

CHAPTER-3

Evaluation of some descriptors of the real chemical world based on electronegativity:

Index

3.2.3.Hetaropolar bond length:

3.2.4.Dipole moment:

3.2.5. Atomic Polar Tensor(APT):

3.2.6. Standard enthalpies of formation and bond dissociation energy:

3.3.Results and discussion:

Conclusion:

Reference:

3.1 Introduction:

In the previous two chapters, we have discussed the concepts, scales and the applications of the electronegativity. In this chapter, for a validity test of the applications of the electronegativity, we have calculated the molecular electronegativity, internuclear distance, atomic polar tensor (APT), bond energy, standard enthalpies of formation, dipole charge, and dipole moment of some extremely ionic compounds and covalent molecules through various ansatz which are well-known in their field and well-accepted by the scientific community.

We have found some dimensional inconsistency in the algorithms for the evaluation of bond energy. We venture to suggest a new semi-empirical ansatz for the evaluation of the bond energy as $BE=a\ (\Delta\chi/R_{AB})+b$ in eV, where $\Delta\chi$ is the electronegativity difference, R_{AB} is the internuclear distance of the constituents forming the bond in Å, and 'a' and 'b' are the constants depending upon the nature of bond between the atoms A and B. We have applied our suggested ansatz to calculate the bond energy of the alkali halide and hydrogen halide molecules.

We also made comparative study of the evaluated bond energy of some selected molecules vis a vis their experimental counterparts and found beautiful correlation between them. Furthermore, we proposed a new formula for the evaluation of standard enthalpies of formation as $-H_f=a\Delta\chi^2+b$. We have performed comparative studies of the data for the standard enthalpies of formation evaluated through the proposed ansatz , $-H_f = a\Delta\chi^2 + b$, of alkali halides vis a vis their experimental counterparts and have found nice correlation. The comparative study of the dipole charge evaluated using different algorithms of alkali halides

The detail study concludes that electronegativity is an intrinsic property of atoms, not in situ property which originates from the attraction of the screened nucleus of the atom on the valence electron. It is important fundamental descriptors which has no quantum mechanical observable, hence empirical. There is no experimental benchmark for electronegativity as it is not experimental determinable quantity. Electronegativity has wide application in the real world of chemical interaction and chemical stability.

3.2.Method of computation

3.2.1Molecular electronegativity:

3.2.1.1. Geometrical mean formula:

Sanderson [1] postulated a geometric mean principle for the electronegativity equalization. He pointed out that a molecule's final electronegativity is the geometric mean of the original atomic electronegativities.

$$\chi GM = (\chi_A \chi_B)^{1/2} \quad (1)$$

Other average principles, such as arithmetic and harmonic avarage found in the literature. We have computed the molecular electronegativity through the atomic electronegativity data computed by Ghosh and Islam[2] invoking the above mention available mathematical averaging technique.

3.2.1.2. The arithmetic mean (AM) formula:

$\chi AM=(\chi_A + \chi_B)/2$ (2)

3.2.1.3. The harmonic mean (HM) formula:

$1/\chi HM=(1/2)\{(1/\chi_A +(1/\chi_B)\}$

Or, $\chi HM=2(\chi_A \chi_B)/ (\chi_A + \chi_B)$ (3)

3.2.1.4. Molecular electronegativity on the basis of Simple Bond Charge(SBC) model:

Ray et al[3] computed the equalized electronegativity invoking the SBC[4] model as follows-

For a diatomic molecule AB with the equilibrium bond length $R_{AB,}$ if we consider Z_A and Z_B as the charge on atom A and B in the diatomic molecule AA having the bond length $2r_A$ and the charge on B in BB having the bond length $2r_B$ respectively and δ is the amount of charge transferred during the process of the molecule formation, then $Z_A + \delta$ and $Z_B - \delta$ is the charges on nuclei A and B in the molecule AB.

Pasternak[5] defined the electronegativity of a bonded atom A in a molecule as

$\chi_A = C(Z_A/r_A)$, (4)

Where, C is the constant which depends on type of bond between A and B.

Thus, for the diatomic AB the natural definition for the electronegativity of atom A and B, in the final molecule after charge transfer, has the form-

$\chi_A(\text{in AB}) = C(Z_A + \delta)/r_1$, (5)

And

$\chi_B(\text{in AB}) = C(Z_B - \delta)/r_2$, (6)

According to Sanderson electronegativity equalization principle [1], electronegativity of bonded atoms in a molecule must be equal to each other.

Thus,

$\chi_{AB} = \chi_A(\text{in AB}) = \chi_B(\text{in AB})$ (7)

$= C(Z_A + \delta)/r_1 = C(Z_B - \delta)/r_2 = C(Z_A + Z_B)/R_{AB}$

or, $\chi_{AB} = (\chi_A r_A + \chi_B r_B)/(r_1 + r_2) = (\chi_A R_{AA} + \chi_B R_{BB})/2R_{AB}$ (8)

3.2.2.Dipole charge:

The most obvious application of electronegativities is the prediction of the polarity of a chemical bond, for which the concept was originally introduced by Pauling[6]. In general, the greater is the difference in electronegativity between two atoms, the more polar is the bond that will be formed between them, with the atom having the higher electronegativity being at the negative end of the dipole.

Pauling[6] proceeded to derive the equation for obtaining the dipole charge, plotted these percentages against their electronegativity differences. Thereafter several empirical ansatzs were suggested by various workers to evaluate the dipole charge invoking electronegativities of the bonded atoms from various scales. The working formulae to evaluate the dipole charge are reviewed below:

3.2.2.1.Pauling's formula :

Pauling [6] proposed an ansatz to calculate the ionic character of the bond (i.e. dipole charge) as-

$$q= 1- \exp(-(\chi_B - \chi_A)^2/4) \quad (9)$$

where χ_B and χ_A are the atomic electronegativities of atoms B and A respectively.

3.2.2.2. Nethercot's formulae:

Nethercot[7] pointed out that the dipole charge,q is not only depends on the electronegativity difference of two atoms but also depends on the equalized electronegativity value. He proposed two formulae to calculate the dipole moment charges as under-

$$q = 1\text{-} \exp(-3(\chi_B - \chi_A)^2/2\,\chi_{AM}{}^2) \qquad (10)$$

$$q = 1\text{-} \exp(-(\chi_B - \chi_A)^{3/2} / \chi_{GM}{}^{3/2}) \qquad (11)$$

where χ_{AM} and χ_{GM} are the arithmetic mean(AM) and the geometric mean(GM) of the two atomic electronegativities.

3.2.2.3. Barbe's formula:

Barbe [8] proposed another simple equation to calculate the dipole moment charges of hetero nuclear diatomic molecules. The ansatz which has been proposed by Barbe to calculate the dipole moment charges is as

$$q = (\chi_B - \chi_A) / \chi_B \qquad (12)$$

where $\chi_B > \chi_A$.

3.2.2.4. Kim's formula:

Using the SBC model and Ray et al electronegativity equalization procedure, Kim [9] calculated the amount of charge transferred, q, for heteronuclear diatomic species as

$$q = \{ r_1 r_2/(r_1 + r_2) \}\{(r_B Z_B/r_2 r_B) - (r_A Z_A/r_1 r_A)\} \qquad (13)$$

Applying the zero-order approximation,i.e. $r_1 \approx r_A$ and $r2 \approx r_B$,Kim[9]obtained

$$q = \{ r_1 r_2/CR_{AB}\}(\chi_B - \chi_A) \qquad (14)$$

The value of C was taken as 6.9696 eVÅ/e

3.2.3.Hetaropolar bond length:

Pauling[6] first evaluated the bond length from the electronegativity. The Pauling electronegativity was derived from heats of formation or essentially, bond energies, the electronegativity difference between two atoms reflects the strength of two bonds and moreover there exist a quantitative correlation between electronegativity and bond polarity. On the basis of simple bond charge model(SBC)[4], Ray et al[3] derived the heteropolar bond length, R_{AB}, in terms of the electronegativites, χ_A and χ_B , and covalent radii , $r_A = 1/2R_{AA}$ and $1/2R_{BB}$, of atoms A and B as follows-

$$R_{AB} = (r_A+r_B)-\{(r_A r_B(\chi^{1/2}{}_A - \chi^{1/2}{}_B)^2\}/ (\chi_A\, r_A + \chi_B\, r_B\,) \qquad (15)$$

3.2.4.Dipole moment:

Dipole moments μ_d are caused by two opposite charges of magnitude q in Coulombs separated by distance r in meters.

$$\mu_d = q \times r \qquad (16)$$

In the molecular world we use the following form for defining the molecular dipole moment

$\mu_d = q \times R_{AB}$. (17)

where R_{AB}, is the internuclear distance.

In Debye,

$\mu_d = 4.8\, q \times R_{AB}$. (18)

where R_{AB} must be expressed in Å unit.

3.2.5. Atomic Polar Tensor(APT):

Kim [9] extended the SBC model to evaluate the Atomic polar tensor. He showed that electronegativity and the electronegativity equalization can be used as an important tool for the determination of Atomic Polar Tensor in case of diatomic molecule. The centroid of positive charger, r, relative to the point defining the centroid of negative charge given by Kim as

$r = \{r_2 Z_B - r_1 Z_A - (r_1 + r_2)\, q\} / (Z_A + Z_B)$ (19)

The dipole moment, μ, was defined by Kim[9] as to direct from the centroid of negative charge to that of positive charge, i.e.,

$\mu = (Z_A + Z_B) r = - (r_1 + r_2)\, q + (r_1 Z_B - r_2 Z_A)$ (20)

$= - R_{AB}\, q + (r_2 Z_B - r_1 Z_A)$

$$= - 1/C\ [(r_A r_B\ \chi_A\ \chi_B) / (r_A\ \chi_A + r_B\ \chi_B)^2][R^2_{AB}\ (\chi_B - \chi_A)] \qquad (21)$$

For a diatomic molecule of AB type where A atom is located at the origin and B atom in positive Cartesian direction and $\chi_B < \chi_A$. Then, the atomic polar tensors(P_x's) for atoms. A and B was given by Kim[9] as

$$P_x^B = - P_x^A = (\partial\mu/\partial R)_e \qquad (22)$$

where $(\partial\mu/\partial R)_e$ is the dipole moment derivative at geometric equilibrium.

The simple bond charge (SBC) model [4] for the dipole moment under this condition gives the atomic polar tensor of B atom as follows,

$$P_x^B = (\partial\mu/\partial R)_e = - (\chi_B - \chi_A).\ 2R_{AB} r_A r_B X_A\ X_A / C(r_A \chi_A + r_B\ \chi_A)^2 \qquad (23)$$

3.2.6. Standard enthalpies of formation and bond dissociation energy:

Bratsch[10] pointed out that the Pauling scale of electronegativity can be used to predict standard enthalpies of formation of binary compounds-

$$\Delta H^0_f = -96.5n\ [\chi_A - \chi_B]^2 \qquad (24)$$

where n is the number of equivalents in the compound formula, χ_A, χ_B are the Pauling electronegativity in $(eV)^{1/2}$ unit. ΔH^0_f is the Kilo joule per mole.

The Bond dissociation energies or the bond energy (BE) for a bond A-B is defined as the standard-state enthalpy change for a reaction of the type AB→A+B at a specified temperature (T).

$(BE)_T=\Delta H_f^0(A)+ \Delta H_f^0(A)- \Delta H_f^0(AB)$ (25)

where ΔH_f^0 is the standard state heat of formation.

The earliest method which correlate the bond energy with electronegativity is the landmark Pauling's equation[6] which is based on single bond energy. The scale correlates the extra ionic resonance energy to the electronegativities of atoms. The energy difference Δ, was defined by-

$\Delta=D(A\text{-}B)-(1/2)[D(A\text{-}A)+D(B\text{-}B)].$ (26)

Pauling was able to assign electronegativity of many elements which roughly satisfy the equation

$\Delta=(\chi_A- \chi_B)^2$ (27)

Pauling took Δ as the negative counterpart of the heat of formation (ΔH_f^0).

The formula for bond enthalpy of Bratsch[10], $\Delta H^0{}_f$ =-96.5n $[\chi_A\text{-}\chi_B]^2$, has some dimensional inconsistency in the algorithms because unless we define the dimension of the constant n as $(\text{energy})^{1/2}$ we can't evaluate the enthalpy in energy unit. Thus in a venture to suggest a new semi-empirical ansatz for the evaluation of the bond energy, we have found a linear correlation between the enthalpy of the bond and the square of the electronegativity difference of the atoms constituting the bond. Thus, we have suggested the ansatz for the evaluation of the bond energy as-

$-H_f = a\Delta\chi^2 + b$ in energy unit. (28)

where a and b are the correlation constants depends upon the nature of bond between the atoms A and B . The dimension of a is $(\text{energy})^{1/2}$ and b is dimension less quantity.

We have also noticed that the use of Pauling's formula in the study of the bond energy of molecules is problematic for any scale other than Pauling, because all modern scales of electronegativity are express in energy unit. So if one use other scales in the Pauling Bond energy formula the dimensional problem arise. Notice on the physical nature of bond, we have suggested a new formula for the evaluation of bond energy as-

$BE = a\,(\Delta\chi/R_{AB}) + b$ in eV, (29)

where $\Delta\chi$ is the electronegativity difference, R_{AB} is the internuclear distance of the constituents forming the bond in Å, and 'a' and 'b' are the constants depends upon the nature of bond between the atoms A and B the dimension of a is Å^{-1} and b is a dimensionless quantity. We have applied our suggested ansatz to calculate the bond energy of the alkali halide and hydrogen halide molecules.

We have used the ansatzs (1) ,(2), (3) and (8) to calculate the molecular electronegativity using the set of atomic electronegativities proposed by Ghosh and Islam[2] very recently. The dipole charge of the heteronuclear diatomic molecules are calculated using the algorithms (9), (10),(11),(12) and (14). The internuclear distances of the alkali halides and hydrogen halides are computed through the equation (15). We have computed the atomic polar tensor of the two sets of compounds using the ansatz (23). The enthalpy of bond formation and the bond energy of the alkali halides are computed through our suggested semi-empirical algorithms (28) and (29) respectively.

The computed results and the comparative studies are presented in the tables as follows-

Table-1: Comparative study of the evaluated molecular electronegativity based on different algorithms of alkali halides

Table-2: Comparative study of the evaluated molecular electronegativity based on different algorithms of hydrogen halides

Table-3: Comparative study of the evaluated internuclear distance of alkali halides vis a vis their spectroscopic counterparts.

Table-4: Comparative study of the evaluated internuclear distance of hydrogen halides vis a vis their spectroscopic counterparts.

Table- 5 Computed ‘a’ and ‘b’ value for the evaluation of bond energy(BE) of formation evaluated through the proposed ansatz, $BE=a(\Delta\chi/R_{AB})+b$, of alkali halides(A-X) and Hydrogen halides (H-X)

Table-6: Comparative study of the evaluated bond energy of alkali halides vis a vis their experimental counterparts.

Table-7: Comparative study of the evaluated bond energy of hydrogen halides vis a vis their experimental counterparts.

Table-8: Computed ‘a’ and ‘b’ value for the evaluation of Standard enthalpies of formation evaluated through the proposed ansatz, $H_f=a(\chi_A-\chi_B)^2+b$, of alkali halides(A-X) and Hydrogen halides (H-X)

Table-9: Comparative study of the Standard enthalpies of formation evaluated through the proposed ansatz , $-H_f=a(\chi_A-\chi_B)^2+b$, of alkali halides vis a vis their experimental counterparts.

Table-10: Comparative study of the evaluated Standard enthalpies of formation evaluated through the proposed ansatz , $-H_f=a(\chi_A-\chi_B)^2+b$, of hydrogen halides vis a vis their experimental counterparts

Table-11: Evaluated Atomic polar tensor (APT) of some selected diatomic heteronuclear molecules.

Table-12: Comparative study of the evaluated Atomic polar tensor (APT), experimental counterparts and Kim evaluated values of some selected diatomic heteronuclear molecules.

Table-13: Comparative study of the dipole charge evaluated using different algorithms of alkali halides

Table-14: Comparative study of the dipole charge evaluated using different algorithms of hydrogen halides

Table-15: Comparative study of the dipole moments (Debye) evaluated using different algorithms of alkali halides vis a vis their experimental counterparts(Debye)

Table-16: Comparative study of the dipole moments (Debye) evaluated using different algorithms of hydrogen halides vis a vis their experimental counterparts(Debye).

To better visualization of the computed results and the comparative studies we have drawn the figures as under-

Figure-1: Comparative plot of the evaluated molecular electronegativity based on different algorithms of alkali halides

Figure-2: Comparative plot of the evaluated molecular electronegativity based on different algorithms of hydrogen halides

Figure-3: Comparative plot of the evaluated internuclear distance of alkali halides vis a vis their spectroscopic counterparts.

Figure-4: Comparative plot of the evaluated internuclear distance of hydrogen halides vis a vis their spectroscopic counterparts.

Figure-5: Comparative plot of the evaluated bond energy of alkali halides vis a vis their experimental counterparts.

Figure-6: Comparative plot of the evaluated bond energy of hydrogen halides vis a vis their experimental counterparts

Figure -7: Comparative plot of the computed Standard enthalpies of formation of alkali halides vis a vis their experimental counterparts.

Figure -8: Comparative plot of the computed Standard enthalpies of formation of hydrogen halides vis a vis their experimental counterparts.

Figure-9: Comparative plot of the evaluated Atomic polar tensor (APT), experimental counterparts and Kim evaluated values of some selected diatomic heteronuclear molecules.

Figure-10: Comparative study of the dipole charge evaluated using different algorithms of alkali halides

Figure-11: Comparative plot of the dipole charge evaluated using different algorithms of hydrogen halides

Figure-12: Comparative plot of the dipole moment evaluated using different algorithms of alkali halides

Figure-13: Comparative plot of the dipole moment evaluated using different algorithms of hydrogen halides

3.3.Results and discussion:

A look on the Tables 1 and 2 and figures 1 and 2 reveals that the molecular electronegativity computed through different algorithms shows similar trends of variation with respect to the nature of the molecules. The trend in variation for a group(M-X, X varies only) not always linear. The electronegativity of M-Br is found to be greater than that of M-Cl. This may be due to the fact that the electronegativity of Br>Cl reported by Ghosh and Islam [2]. This anomalous trend may be eliminated by the proper choice of the electronegativity value. Here it is important to mention that Ghosh and Islam [2] reported the electronegativity value for 103 elements of the periodic table in order to justify their statement that in case of atoms the electronegativity and the hardness are originated from the same fundamental source-the electron attraction power of the screened nucleus. They wanted to verify their statement that the electronegativity and the hardness are fundamentally as well as operationally same quantity.

A look on the Tables 3 and 4 and figures 3 and 4 reveals that there are beautiful correlation between the evaluated internuclear distances computed through the semi-empirical ansatz of Ray et al and the electronegativity value reported by Ghosh and Islam[2]. A careful analysis of the Tables 6 and 7 and figures 5 and 6 reveals that the bond energy computed through our proposed ansatz correlated excellently with their experimentally reported values [11]. This correlation support that the present method of evaluation of the bond energy can be used as a tool for the computation of the bond energies of the heteronuclear diatomic molecules. A look on the Tables 9 and 10 and figures 7 and 8 reveals that enthalpy data computed through our proposed ansatz correlated excellently with their experimentally reported values[12]. Table 12 and figure 9 demonstrates that in maximum case the atomic polar tensor values of the present calculation using the electronegativity value of Ghosh and Islam are more consistent with their experimental[9] counterpart than the values reported by Kim[9]. From Tables 13 and 14 and figures 10 and 11 we see that the dipole charges evaluated through all the formulae are found to be consistent with the known chemico-physical characteristics of such compounds. A deeper scrutiny of computed dipoles of series of binary compounds- alkali metal dimers and hydrogen halides reveals that the dipoles are just consistent with the known physical behavior of such dimers. It is expected that the elements of small electronegativity difference will form covalent bonds and the partial charge densities on atomic sites will also be small and the bond component will have small contribution to the total dipole. From the Tables 15 and 16 and figures 12 and 13 it is distinct that the calculated dipole moments through Nethercot's formulae (AM and GM) of the alkali halides show a nice correlation with the experimental dipole moments. Overall study indicates that Nethercot's arithmetic mean average formula is the best for the evaluation of molecular dipole moment of alkali halides. The computed dipoles moment using the Pauling and Kim's formula are found to be inconsistence with experimental data. The value computed using the electronegativity values of Ghosh and Islam and the formulae of Nethercot and Barbe beautifully correlate our chemical experiences. The computed values predict that the compounds reported in Table 15 are grossly ionic and hence the bond moments are expected to be high. We have presented the computed dipole moments of hydrogen halides along with the available experimental value in Table 16. From our chemical experience we can infer that the hydrogen halides are predominantly covalent in nature. The computed dipoles charges (Table 14) are accordingly less. The Table 16 demonstrates that the experimental dipoles of hydrogen halides are small. The computed dipoles, though scattered, fairly correlate with theoretical dipoles. We feel it necessary to refer to the fact that most likely the lone pair component [13] of the dipoles of such molecules must vectorially couple with the bond moment component. It, therefore, transpires that there can be no good correlation between the experimental dipoles having two contributing components and the bond dipoles of molecules.

Conclusion:

In absence of any theoretical or experimental bench mark to perform the validity test of any scale of electronegativity, we have computed the electronegativity dependent descriptors viz, Internuclear distance, Bond energy, standard enthalpy, atomic polar tensor, charge densities on atomic sites and dipole moments of two sets of widely different heteronuclear diatomic molecules invoking available ansatz for this purpose and using the electronegativities we have just use the ansatz of Ghosh and Islam. We have allayed the seemingly dimensional mismatch in all previous calculations and evaluated the bond energy and standard enthalpy of bond formation in proper energy unit through our suggested algorithms. We have invoked five different formulae to calculate the dipole charges and hence the bond dipole moments. The comparative analysis of dipole moments evaluated through different ansatz and experimental values reveal that the ansatz proposed by Nethercot to compute the dipole charges on atomic sites is much useful as compared to the other methods. The dipole charges are quite consistent with the known chemico-physical nature of the compounds brought under investigation. The computed dipole moments are realistic descriptors of the charge distribution in the molecules under study. The nature of variation of bonds from strongly covalent to strongly ionic is well reflected in the dipole charges and also in the evaluated dipole moments. The correlation between the theoretical dipole moments and the available experimental dipole moments are satisfactory. The validity test of a scale of electronegativity is successful in terms of the computed descriptors of the molecular world under investigation.

Reference:

[1] R. T. Sanderson, Science, 114, (1951) 670.

[2] D.C. Ghosh, N. Islam, Int. J. Quantum Chem. 109,(2009), in press.

[3] N.K. Ray, L. Samuels, R.G. Parr, J Chem Phys70, (1979), 3680.

[4] (a) R.G.Parr, R.F. Borkman, J Chem Phys 46, (1967), 3683, (b)R.F. Borkman, R.G. Parr ,J Chem Phys 48, (1968),1116, (c) R.G. Parr, R.F. Borkman, J Chem Phys 49, (1968),1055.

[5] A.Pasternak, Chem.Phys., 26, (1977), 101.

[6] L. Pauling, The Nature of the Chemical Bond.3rd edn.;Cornell University:Ithaca.NY .(1960), (b) ibid, J. Am. Chem. Soc., 54, (1932), 3570. (c) L. Pauling, D.M. Yost, Proc. Nat. Acad. Sci., 18, (1932),414.

[7] (a) A.H. Nethercot, Jr. Chem Phys Lett 59: (1978),346, (b) ibid, Chem Phys 59, (1981),297.

[8] J.Barbe, J Chem Educ 60, (1983)640.

[9] K.Kim, Bull K Chem Soc 8, (1987),432.

[10] S.G. Bratsch, J Chem Edu,65,(1988),877.

[11] J.E.Huheey,E.A. Keiter,R.L. Keiter, Inorganic Chemistry,Addision-Wesley Publishing Company,N.Y. 4th edn

[12] D.W.Ball, Thomson Brooks/Cole, UK, 1st edn.

[13] DC, Ghosh S Bhattacharyya Int J Quant Chem 105(2005) 270

Tables:

Table-1: Comparative study of the evaluated molecular electronegativity based on different algorithms of alkali halides

Alkali halides	(χAB)AM	(χAB)GM	(χAB)HM	(χAB)Ray et al
LiF	5.74055	4.650192996	3.76693782	5.095936352
LiCl	4.1204	3.732275247	3.380710251	4.06425745
LiBr	4.14285	3.746531471	3.388126063	4.214002746
LiI	3.97925	3.64136361	3.332167856	4.155781668
NaF	5.7753	4.717753348	3.853859825	4.876983958
NaCl	4.15515	3.786499626	3.450556399	3.961812017
NaBr	4.1776	3.800962971	3.45828215	4.106402425
NaI	4.014	3.694267179	3.400002489	4.059773643
KF	5.7169	4.603646104	3.70717652	4.456824049
KCl	4.09675	3.694916408	3.332497043	3.703750562
KBr	4.1192	3.709029931	3.339702619	3.846479994
KI	3.9556	3.604914766	3.285319666	3.820921387
RbF	5.61345	4.39424881	3.439849397	4.246178398
RbCl	3.9933	3.526852773	3.114890061	3.536874697

RbBr	4.01575	3.540324341	3.121184446	3.685533266
RbI	3.85215	3.440944864	3.073634609	3.670294914
CsF	4.8947	2.493757977	1.270522984	3.709708868
CsCl	3.27455	2.001506428	1.223382749	2.934584418
CsBr	3.297	2.00915161	1.224352499	3.150831775
CsI	3.1334	1.952753264	1.216967291	3.171267294
FrF	5.04735	2.999840546	1.782924366	3.749090833
FrCl	3.4272	2.4076916	1.69146208	3.013895661
FrBr	3.44965	2.416888293	1.693316429	3.213437133
FrI	3.28605	2.349044482	1.679222769	3.228549695

Table-2: Comparative study of the evaluated molecular electronegativity based on different algorithms of hydrogen halides

Hydrogen halides	$(\chi_{AB})_{AM}$	$(\chi_{AB})_{GM}$	$(\chi_{AB})_{HM}$	(χ_{AB})Ray et al
HF	7.7682	7.65205099	7.537638623	8.266248204
HCl	6.14805	6.141586064	6.135128924	6.00628647
HBr	6.1705	6.165045165	6.159595153	6.026675382
HI	6.0069	5.991987868	5.977112755	5.752637593

Table-3: Comparative study of the evaluated internuclear distance of alkali halides vis a vis their spectroscopic counterparts.

Alkali halides	R_{AB} (Å)	R_{spect}(Å)	Alkali halides	R_{AB} (Å)	R_{spect}(Å)
LiF	1.76344	1.56388	RbF	2.421236	2.27036
LiCl	2.206001	2.02069	RbCl	2.900973	2.78677
LiBr	2.348568	2.02149	RbBr	3.036601	2.94478
LiI	2.547067	2.39194	RbI	3.236641	3.17692
NaF	1.971816	1.92597	CsF	1.998132	
NaCl	2.422097	2.36082	CsCl	2.51888	
NaBr	2.56356	2.50207	CsBr	2.641519	
NaI	2.762455	2.39194	CsI	2.841424	
KF	2.331182	2.17148	FrF	2.142345	
KCl	2.799607	2.66668	FrCl	2.658096	
KBr	2.937793	2.82081	FrBr	2.782796	
KI	3.137488	3.04788	FrI	2.982848	

Table-4: Comparative study of the evaluated internuclear distance of hydrogen halides vis a vis their spectroscopic counterparts.

Hydrogen halides	R_{AB} (Å)	R_{spect}(Å)	Hydrogen halides	R_{AB} (Å)	R_{spect}(Å)
HF	0.953967	0.91682	HBr	1.459548	1.41466
HCl	1.309479	1.27457	Hi	1.648662	1.60913

Table- 5: Computed 'a' and 'b' value for the evaluation of bond energy(BE) of formation evaluated through the proposed ansatz, $BE=a(\Delta\chi/R_{AB})+b$, of alkali halides(A-X) and Hydrogen halides (H-X)

Molecule	a	b
A-X {A=alkali metal, X=Halogen}	0.709	3.029
H-X X=Halogen	0.871	3.379

Table-6: Comparative study of the evaluated bond energy of alkali halides vis a vis their experimental counterparts.

Alkali halides	BE calculated(eV)	BE expt(eV)	Alkali halides	BE calculated(eV)	BE expt(eV)
LiF	5.735594	5.9382	RbF	5.074709	5.078
LiCl	4.151186	4.8086	RbCl	3.944476	4.6013
LiBr	4.09662	4.3319	RbBr	3.914071	3.9899
LiI	3.922339	3.5961	RbI	3.787694	3.4303
NaF	5.42458	4.9433	CsF	6.017958	5.202377
NaCl	4.030723	4.2179	CsCl	4.487966	4.508036
NaBr	3.987863	3.7515	CsBr	4.432281	4.314242
NaI	3.834848	3.1504	CsI	4.251911	3.471706
KF	5.090809	5.078	FrF	5.715717	
KCl	3.925226	4.3837	FrCl	4.33012	
KBr	3.893906	3.9246	FrBr	4.283255	
KI	3.764917	3.3784	FrI	4.121362	

Table-7: Comparative study of the evaluated bond energy of hydrogen halides vis a vis their experimental counterparts.

Hydrogen halides	BE calculated(eV)	BE expt(eV)
HF	5.822815	5.8553
HCl	3.753945	4.4357
HBr	3.688599	3.7546
HI	3.825948	3.053

Table8: Computed 'a' and 'b' value for the evaluation of Standard enthalpies of formation evaluated through the proposed ansatz, $H_f=a(\chi_A-\chi_B)^2+b$, of alkali halides(A-X) and Hydrogen halides (H-X)

Molecule	a	b
A-X {A=alkali metal, X=Halogen}	0.071	2.865
H-X X=Halogen	0.337	0.396

Table9: Comparative study of the Standard enthalpies of formation evaluated through the proposed ansatz , $-H_f=a(\chi_A-\chi_B)^2+b$, of alkali halides vis a vis their experimental counterparts.

Molecule	$-H_f$ Calculated(eV)	$-H_f$ expt (eV)	Molecule	$-H_f$ Calculated(eV)	$-H_f$ expt (eV)
LiF	6.08261191	6.3837928	RbF	6.330197118	-----
LiCl	3.73058021	4.231024491	RbCl	3.861202252	-----
LiBr	3.75298509	3.63959096	RbBr	3.885227861	-----
LiI	3.596272061	2.80223632	RbI	3.71670409	-----
NaF	6.016517737	5.97547878	CsF	7.902949624	-----
NaCl	3.696464557	3.99919747	CsCl	4.772528521	-----
NaBr	3.718426319	3.74218763	CsBr	4.805719342	-----
NaI	3.564942427	2.98255774	CsI	4.570405871	-----
KF	6.127986237	5.87910009	FrF	7.544382437	-----
KCl	3.754190738	4.52358045	FrCl	4.554436764	-----
KBr	3.776897193	4.08106754	FrBr	4.585681053	-----
KI	3.617986493	3.39812607	FrI	4.364552553	-----

Table-10: Comparative study of the evaluated Standard enthalpies of formation evaluated through the proposed ansatz , $-H_f=a(\chi_A-\chi_B)^2+b$, of hydrogen halides vis a vis their experimental counterparts.

Molecule	$-H_f$ Calculated(eV)	$-H_f$ expt (eV)
HF	2.83228989	2.810331208
HCl	0.956636223	0.503084342
HBr	0.376084157	0.486704709
HI	0.27462745	0.637196292

Table-11: Evaluated Atomic polar tensor (APT) of some selected diatomic heteronuclear molecules.

Molecule	APT	Molecule	APT	Molecule	APT
LiF	1.15228	RbF	1.963152	HF	0.188933
LiCl	0.749156	RbCl	1.263236	HCl	-0.05475
LiBr	0.739497	RbBr	1.244472	HBr	-0.05057
LiI	0.684642	RbI	1.15135	Hi	-0.08674
NaF	1.408302	CsF	1.032073	LiH	-0.68303
NaCl	0.890182	CsCl	0.802698	NaH	-0.69334
NaBr	0.880906	CsBr	0.758182	KH	-0.77879
NaI	0.812762	CsI	0.708377		
KF	1.854634	FrF	1.348423		
KCl	1.164583	FrCl	1.00767		
KBr	1.152305	FrBr	0.960914		
KI	1.063141	FrI	0.896713		

Table-12: Comparative study of the evaluated Atomic polar tensor (APT), experimental counterparts and Kim evaluated values of some selected diatomic heteronuclear molecules.

Molecule	APT present calculation	APT expt	APT Kim	Molecule	APT present calculation	APT expt	APT Kim
LiF	1.152279573	0.908	0.901	HBr	-0.05057106	-0.1	-0.036
LiCl	0.749155706	1.001	0.731	Hi	-0.086744178	-0.011	-0.033
NaCl	0.890182205	0.949	0.836	LiH	-0.683029542	0.473	0.467
KCl	1.164583193	1.074	1.047	NaH	-0.693335079	0.541	0.531
HF	0.18893289	-0.317	-0.195	KH	-0.778785251		
HCl	-0.054751666	-0.193	-0.083				

Table-13: Comparative study of the dipole charge evaluated using different algorithms of alkali halides

Alkali halides	q(Pauling)	q(AM)	q(GM)	q (Barbe)	q (Kim)
LIF	0.999987988	0.872903875	0.824796469	0.739241201	0.466230479
LiCl	0.952537609	0.659422818	0.595397125	0.595206437	0.299018303
LiBr	0.956138049	0.66480752	0.600323342	0.5982812	0.327581336
LiI	0.923838685	0.623065776	0.562818747	0.574741668	0.319790374
NaF	0.999984841	0.864149163	0.813289882	0.73160929	0.480914865
NaCl	0.946479585	0.638476038	0.576489113	0.583358904	0.311071532
NaBr	0.950462348	0.644100479	0.581527755	0.586523659	0.342877553
NaI	0.914955958	0.600596871	0.543213344	0.56229517	0.336188018
KF	0.999989762	0.878670425	0.83253146	0.744435293	0.523396104
KCl	0.956323859	0.673493759	0.608330325	0.603269578	0.351928704
KBr	0.959679918	0.678709901	0.613175678	0.606283095	0.391089068
KI	0.92944487	0.638215279	0.57625617	0.58321245	0.388223886
RbF	0.999994977	0.902048006	0.865284278	0.767155329	0.556402525
RbCl	0.970035874	0.73281811	0.66531005	0.638539429	0.385165625
RbBr	0.972466494	0.737258738	0.669760019	0.64128504	0.428793278
RbI	0.95016103	0.702575116	0.635725831	0.620265406	0.428828576
CsF	0.99999998	0.988234373	0.997987127	0.925009609	0.898118696
CsCl	0.998789413	0.976678511	0.984508827	0.883587331	0.678131927

CsBr	0.998922939	0.97699121	0.984969578	0.884471587	0.751075232
CsI	0.997533488	0.974515547	0.981241121	0.877701965	0.76362544
FrF	0.99999993	0.979360774	0.988344197	0.891484105	0.775983739
FrCl	0.997390824	0.952105608	0.944076759	0.831543418	0.581303997
FrBr	0.997662647	0.952867999	0.945361191	0.832822994	0.645269936
FrI	0.994908153	0.946811149	0.935201298	0.823026917	0.655645625

Table-14: Comparative study of the dipole charge evaluated using different algorithms of hydrogen halides

Hydrogen halides	q(Pauling)	q(AM)	q(GM)	q (Barbe)	q (Kim)
HF	0.833214528	0.163125306	0.186878877	0.293921924	0.082446499
HCl	0.07636603	0.012530754	0.02742376	0.096092871	0.019567116
HBr	0.065074433	0.010547511	0.024116	0.087767082	0.018604952
HI	0.163834735	0.029314765	0.051669	0.151507011	0.031335202

Table-15: Comparative study of the dipole moments (Debye) evaluated using different algorithms of alkali halides vis a vis their experimental counterparts (Debye)

Alkali halides	Dipole Moment Pauling	Dipole Moment (AM)	Dipole Moment (GM)	Dipole Moment Kim	Dipole Moment Barbe	Dipole Moment Expt
LIF	8.464412	7.388706	6.981501	3.946414	6.257316997	6.32736
LiCl	10.08623	6.982499	6.304544	3.166246	6.302524615	7.1289
LiBr	10.77867	7.494459	6.767521	3.692866	6.744499777	7.268
LiI	11.29478	7.617553	6.880978	3.909732	7.026746394	7.4285
NaF	9.464572	8.178926	7.697557	4.551722	6.924473698	8.1558
NaCl	11.00383	7.422963	6.702299	3.616538	6.782167882	9.0012
NaBr	11.69552	7.925713	7.15575	4.219138	7.217225172	9.1183
NaI	12.13212	7.963785	7.202892	4.457781	7.455912965	9.2357
KF	11.18956	9.832035	9.315755	5.856631	8.329987399	8.5926
KCl	12.85119	9.050484	8.174811	4.729257	8.106804043	10.269
KBr	13.53284	9.570763	8.646639	5.514905	8.549443633	10.6278
KI	13.99738	9.611484	8.678384	5.846629	8.783144926	10.82
RbF	11.62187	10.48354	10.05628	6.466473	8.915827635	8.5465
RbCl	13.50743	10.20425	9.264223	5.363305	8.891451359	10.51
RbBr	14.17437	10.74605	9.762211	6.249956	9.347168784	10.86
RbI	14.76159	10.91512	9.876559	6.662228	9.63636749	11.48

CsF	9.591032	9.478188	9.571727	8.613885	8.871796785	7.8839
CsCl	12.07599	11.80865	11.90333	8.199038	10.68312223	
CsBr	12.66563	12.38756	12.48872	9.523101	11.2144727	10.82
CsI	13.6052	13.29126	13.38299	10.41496	11.9708336	11.69
FrF	10.28326	10.07102	10.1634	7.97964	9.167360289	
FrCl	12.72557	12.14778	12.04534	7.416777	10.60954743	
FrBr	13.3262	12.72786	12.62759	8.619144	11.12436906	
FrI	14.24477	13.55613	13.3899	9.387319	11.78382894	

Table-16: Comparative study of the dipole moments (Debye) evaluated using different algorithms of hydrogen halides vis a vis their experimental counterparts (Debye).

Hydrogen halides	Dipole Moment Pauling	Dipole Moment (AM)	Dipole Moment (GM)	Dipole Moment Kim	Dipole Moment Barbe	Dipole Moment Expt
HF	3.815324	0.746958	0.855726	0.377526	1.345880717	1.8265
HCl	0.479999	0.078762	0.172372	-0.12299	-0.603991664	1.1086
HBr	0.4559	0.073894	0.168953	-0.13034	-0.614881293	0.828
HI	1.296519	0.231985	0.408887	-0.24797	-1.19896249	0.4477

Figures:

Figure-1: Comparative plot of the evaluated molecular electronegativity based on different algorithms of alkali halides

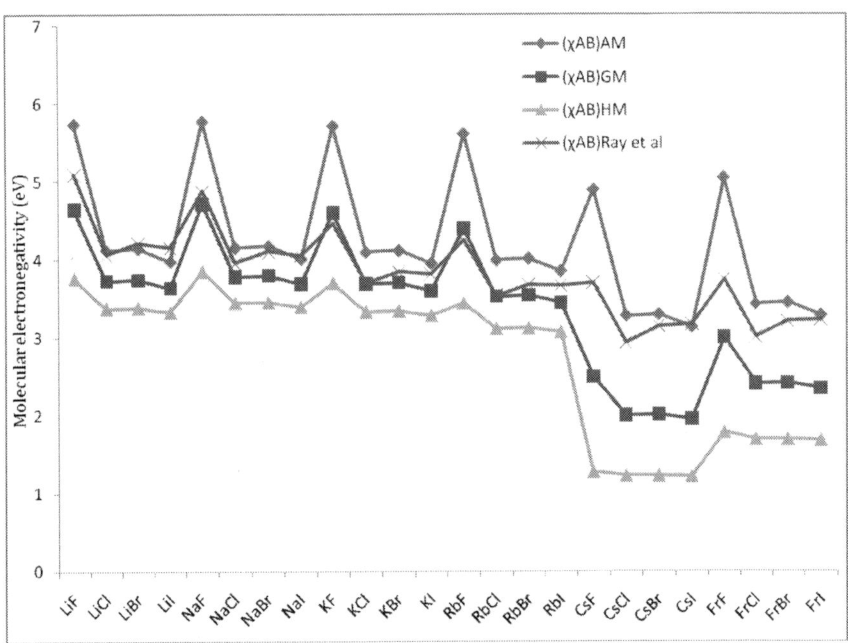

Figure-2: Comparative plot of the evaluated molecular electronegativity based on different algorithms of hydrogen halides

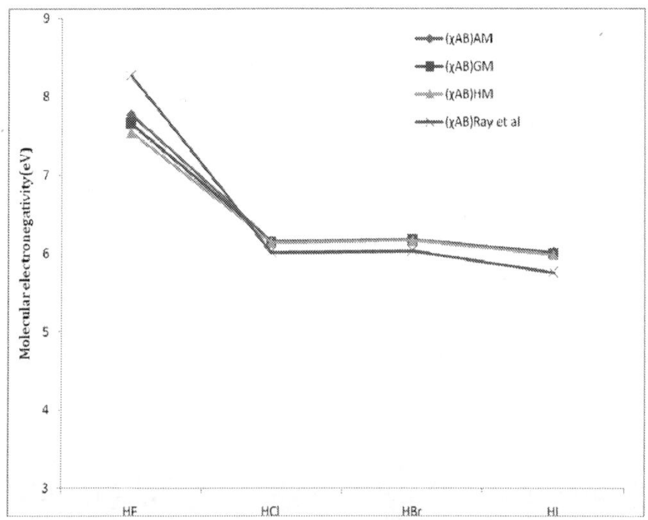

Figure-3: Comparative plot of the evaluated internuclear distance of alkali halides vis a vis their spectroscopic counterparts.

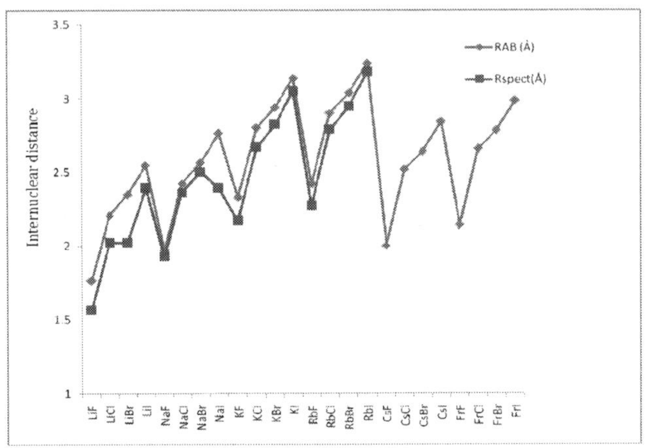

Figure-4: Comparative plot of the evaluated internuclear distance of hydrogen halides vis a vis their spectroscopic counterparts.

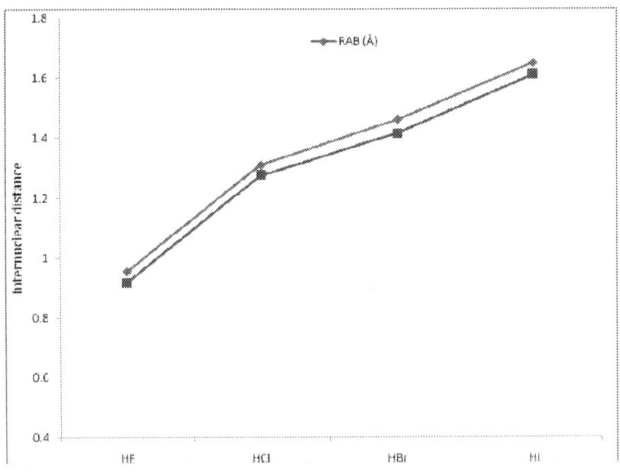

Figure-5: Comparative plot of the evaluated bond energy of alkali halides vis a vis their experimental counterparts.

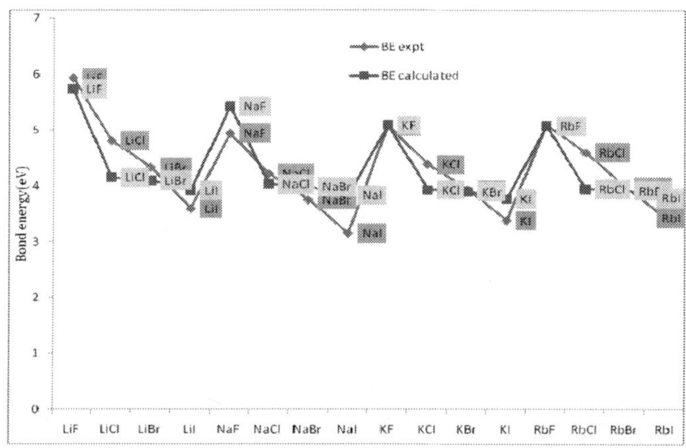

Figure-6: Comparative plot of the evaluated bond energy of hydrogen halides vis a vis their experimental counterparts

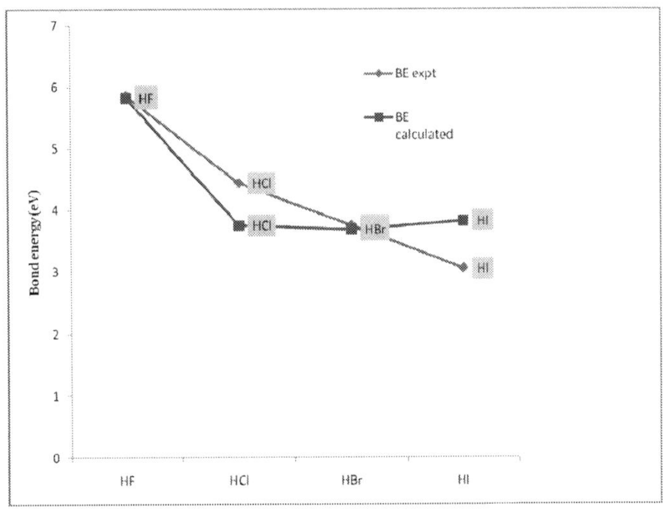

Figure -7: Comparative plot of the computed Standard enthalpies of formation of alkali halides vis a vis their experimental counterparts.

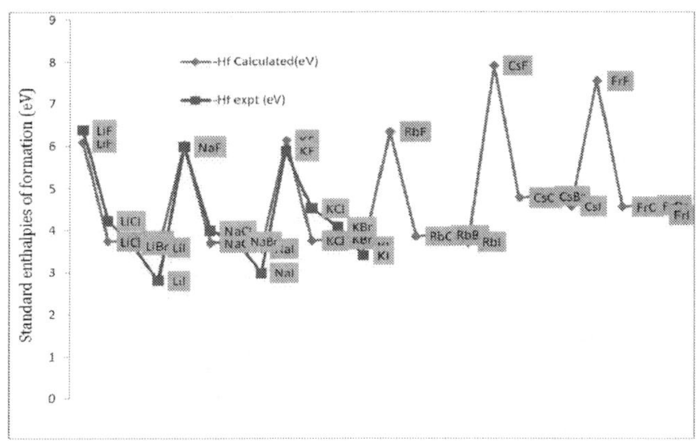

Figure -8: Comparative plot of the computed Standard enthalpies of formation of hydrogen halides vis a vis their experimental counterparts.

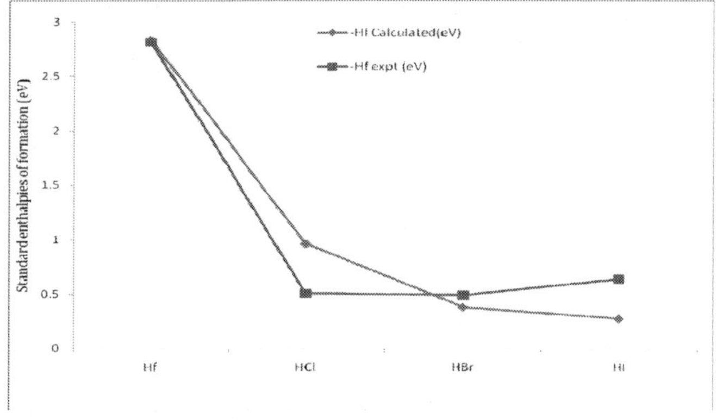

Figure-9: Comparative plot of the evaluated Atomic polar tensor (APT), experimental counterparts and Kim evaluated values of some selected diatomic heteronuclear molecules.

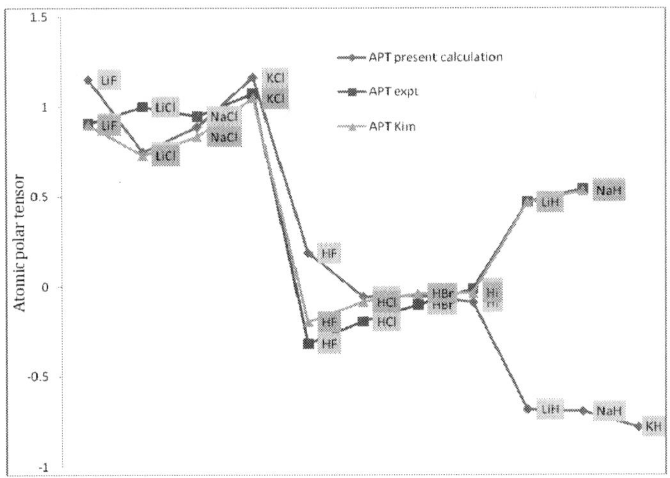

Figure-10: Comparative study of the dipole charge evaluated using different algorithms of alkali halides

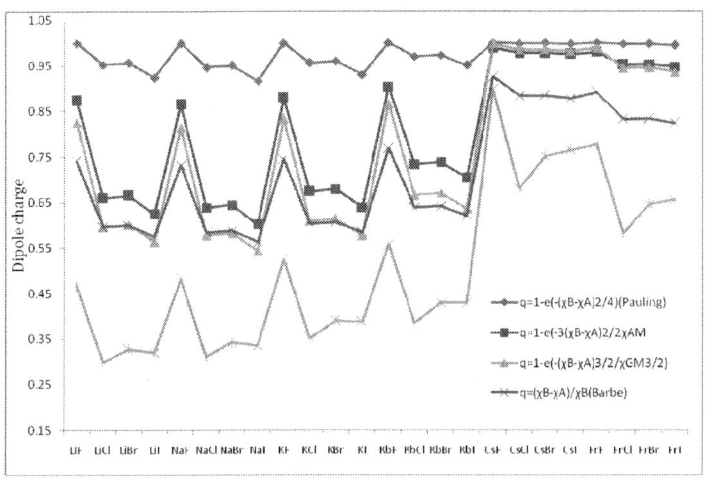

Figure-11: Comparative plot of the dipole charge evaluated using different algorithms of hydrogen halides

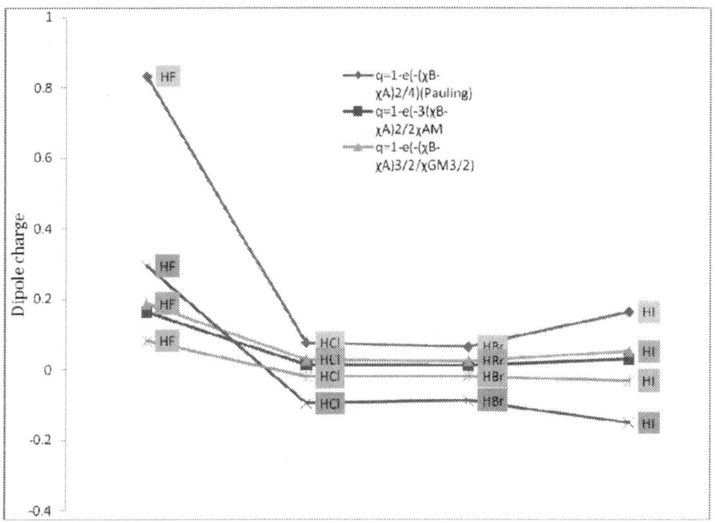

Figure-12: Comparative plot of the dipole moment evaluated using different algorithms of alkali halides

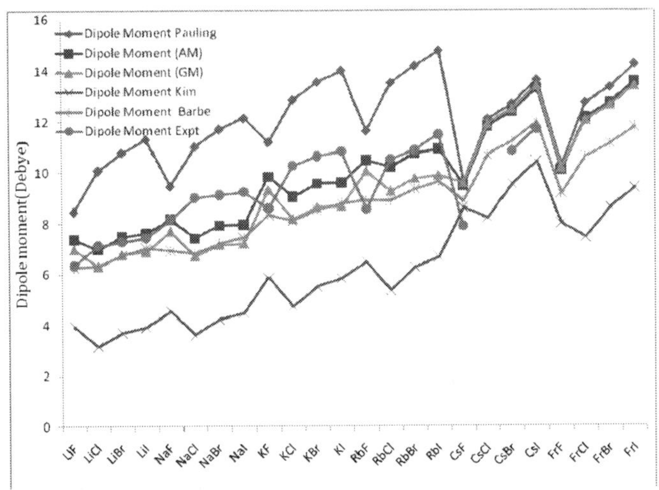

Figure-13: Comparative plot of the dipole moment evaluated using different algorithms of hydrogen halides

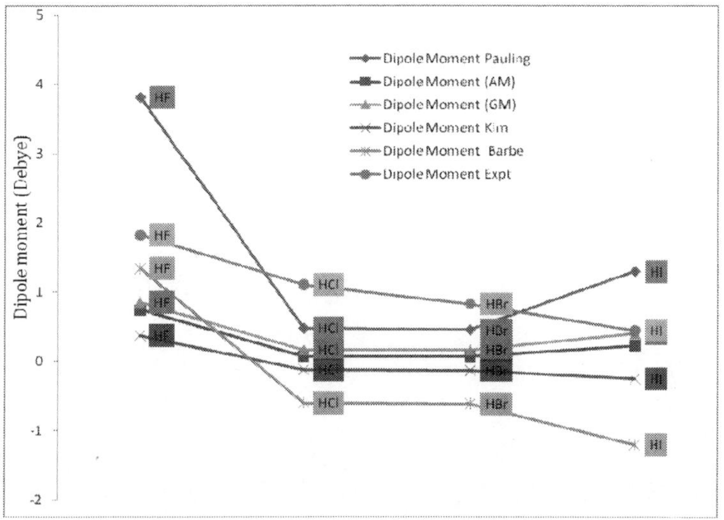

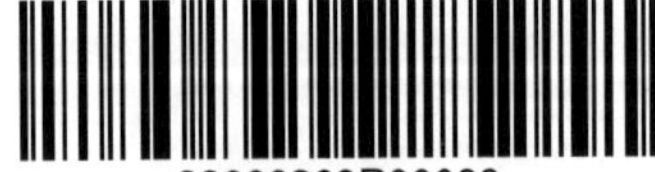

Made in the USA
Middletown, DE
22 December 2016